5/81

DIGITAL IC's —
How They Work
and
How to Use Them

DIGITAL IC's —
How They Work
and
How to Use Them

Alfred W. Barber

Parker Publishing Company, Inc.
West Nyack, N.Y.

Originally Published as

Practical Guide to Digital Integrated Circuits

Reward Edition . . . . . . November 1980

Library of Congress Cataloging in Publication Data

Barber, Alfred W
    Practical guide to digital integrated circuits.

    Bibliography: p.
    Includes index.
    1. Digital integrated circuits. I. Title.
TK7874.B37   621.381'73   76-10138

How This Book Will Help You

This book provides *practical* guidance in the understanding, use and application of integrated circuits. Theory alone cannot anticipate what happens when incredibly efficient devices are combined and put to real use. Manufacturers' specifications and application advice leave many gaps which can be filled by long experience but which this book will help to fill more quickly and easily. The emphasis is on understanding what goes on in each device and in the final systems.

A long and varied experience in many areas of electronic circuit design has convinced me that one can understand integrated circuits and their applications best by understanding their actual or equivalent circuitry, so this book will place considerable emphasis on basic internal circuit performance. All references to the actual circuit performance of a given integrated circuit have been verified by extensive operational tests. Several special test devices have been developed to expedite these tests. (For example, the section entitled "Simple IC testers for simple IC devices" in Chapter 10 describes how to make these testers.)

Chapter 1 discusses the logical development of integrated circuits; what they are; how they came into being; how they are manufactured; and how they are classified.

The basic building block of digital integrated circuits is the "gate." This is a basic switch which has two states, ON and OFF, and which when multiplied and combined provides most of the essential characteristics in digital integrated circuits.

This book will give you a functional understanding along with many practical uses for ICs. Chapter 2 introduces the concept of logic to explain "gate operation" (for instance, one section explains the eight permutations of basic gates with equations and equivalencies); Chapter 3 describes many basic forms of logical ICs (gates) from the standpoint of internal circuitry; Chapter 4 deals with the various forms of flip-flops and shows many of their practical uses; and Chapter 5 shows how combining flip-flops provides countless useful circuits.

Chapter 6 offers "A Key to Effective Circuit Design Is Breadboarding" and Chapter 9 explains "How to Specify and Select ICs." Every chapter is based on the day-to-day experiences of those who put integrated circuits to use in a large variety of actual circuits and systems.

When integrated circuits are combined to form systems, many practical problems arise which can often be avoided by reviewing Chapter 7: "How to Devise Complex Systems by Using Simple Basic Elements."

Chapter 11 will show you "How to Use ICs in Hobby Projects." For example, "how to invent a new hobby device from start to finish" is described in detail, based on an original development carried out by the author. This device is intended primarily to control electrical equipment, in this case model electric trains, by means of voice commands.

The extensive bibliography identifies a large number of references in the field of digital ICs for further reading. A practical glossary of terms used in the IC field is also included.

It is characteristic for some semiconductor manufacturers to develop and favor a particular type of logic (RTL, TTL, C/MOS and so on). This characteristic has carried over to many publications, including books on the subject. Technicians and others have often found it misleading and frustrating to read such a book since it may present a rather narrow view of the subject. It is also unfair to the engineer who may be led to believe such a narrow treatment is representative of the state-of-the-art when, in fact, it represents only a small segment. This book is both impartial and unbiased in this regard. You will find it a helpful, realistic, and indeed, a very practical guide to digital integrated circuits.

Alfred W. Barber

Table of Contents

HOW THIS BOOK WILL HELP YOU 7

Chapter 1

UNDERSTANDING THE RELATIONSHIP BETWEEN
INTEGRATED CIRCUITS AND TRANSISTORS 19

    Putting more than one transistor on a substrate started
        it all   19
    Actual and equivalent circuits aid in understanding ICs
        21
    A whole new technology was developed to manufac-
        ture ICs   22
    Field effect ICs have some advantages   24
    Defining small, medium and large scale integration
        26
    Several basic packaging methods and forms were de-
        veloped   27

Chapter 2

USING LOGIC TO EXPLAIN THE OPERATION OF ICs 28

    Symbols adopted for greatest usefulness   29
    How simple ICs provide the basic logic gates   31
    Truth tables are one practical aid in analyzing circuits
        31

Why inverting gates require fewer transistors than
non-inverting gates   36
Negative logic helps in understanding inverting ICs
38
The exclusive OR gate is a very useful device   39
Eight permutations of basic gates with equations and
equivalences   40

Chapter 3

UNDERSTANDING HOW CIRCUITS HAVE BEEN
USED IN LOGICAL ICs 42

Logic ICs have evolved from earlier vacuum tube cir-
cuits   42
How transistors are used to provide the three major
building blocks   43
Many logic types have been developed, including
RTL-RCTL-DTL-TTL-ECL-HTL-UTL-
MOS/FET-C/MOS   43
Wired-AND/OR gates are available because of transis-
tors   59
Complex logic circuits are built from simple basic
gates   60

Chapter 4

HOW FLIP-FLOPS ARE FORMED AND USED 61

Defining the flip-flop   61
Synchronous flip-flop operation is required in some
cases   64
The toggle flip-flop (R-S-T) is one simple form   66
The J-K flip-flop has the most versatile capabilities
66
Other flip-flops have been devised   67
Some practical ways to use flip-flops   68
Pulses and synchronous operation as applied to flip-
flops   68
Clocking the system a "must" in many applications
71

Chapter 5

COMBINING GATES AND FLIP-FLOPS FOR A NEW
WORLD OF CAPABILITIES 74

How serial counters are formed of cascaded flip-flops
        74
Counters can divide frequency by almost any factor
        74
Counters are used to provide interval timers   78
Flip-flops can be interconnected to form shift registers
        78
Flip-flops connected another way become ring count-
        ers   79
Large numbers of flip-flops are interconnected to form
        random-access memory (RAM)   79
Permanent internal connections provide read-only-
        memories (ROM)   84
Multiplexers are formed of gates   86
Encoders and decoders are also formed of gates   87
Code converters are very useful combinations of
        gates, too   88
The data selector is a useful gating concept   90
Adders are basic to logic computers   90
Analog-to-digital converters are an interface between
        the real world and the world of ICs   92
Signal conditioners for A/D converters can improve
        accuracy   95
Digital-to-analog conversion back to the real world
        97

Chapter 6

A KEY TO EFFECTIVE CIRCUIT DESIGN
IS BREADBOARDING 102

Defining breadboarding   102
Why one should do breadboarding   102
Many useful devices are available   103
Describing an experiment to develop an exclusive-OR
        circuit   104

Breadboard laboratories provide very complete facilities   106
Breadboard laboratory   107
Simple but interesting circuits are easily set up with experimenter's kits   108
One needs several auxiliary devices to breadboard properly   108
How to solve some breadboarding problems   109
A versatile timing circuit that can be very useful   110

Chapter 7

HOW TO DEVISE COMPLEX SYSTEMS BY USING SIMPLE BASIC ELEMENTS 114

Practical ways to convert block diagrams to hardware   114
Compatibility of components should be examined   114
Noise immunity may spell the difference between success or failure of a system   115
Understanding interface devices and circuits   117
DTL or TTL driving C/MOS, for example   117
C/MOS driving DTL or TTL, another example   118
Interfacing HTL with C/MOS, a further example   119
Interfacing ECL with C/MOS, a still further example   120
The how and why of input/output devices   121
Meeting fan-in and fan-out requirements   121
Using fan-in gate expanders   124
Solving problems with opto-coupling   126
Using the important tool of charting to simplify systems   130

Chapter 8

ANALYZING THE FANTASTIC DEVELOPMENTS IN LARGE SCALE INTEGRATION (LSI) 136

LSI has created a revolution in the IC field   136
Ion implantation is a powerful tool   137

What goes into a pocket calculator   137
How technology is providing high capability pocket
        calculators   140
Electric clocks—use ICs   140
Electronic watches are big users of ICs   143
LSI RAMs are a challenge to LSI technology   145

Chapter 9

HOW TO SPECIFY AND SELECT ICS 148

Factors determining the selection of ICs   148
Some specifications are more important than others
        152
ICs are not alike, making selection important   152
Aplication notes can help provide applications instruc-
        tions   153
Special testing is sometimes called for   153

Chapter 10

SIMPLE AND SOPHISTICATED WAYS OF TESTING
DIGITAL INTEGRATED CIRCUITS 156

How digital IC testing differs from analog circuit test-
        ing   156
How to test to specifications   165
Simple IC testers for simple IC devices   170
Logic probes give more information   172
Logic clips examine several circuits simultaneously
        173
When and how to use logic analyzers   176
Why logic state analyzers are required in some cases
        178
Practical information about automatic testers   184
Automatic test systems (ATS) and automatic test
        equipment (ATE): functionally ideal system;
        hardware realities; software assistance   185
Recommendations for users of ATS and ATE   186

Chapter 11

HOW TO USE ICs IN HOBBY PROJECTS 188

    ICs in the hobby field add potential and interest   188
    Tips on building hobby projects   189
    How to invent a new hobby device from start to finish
        193
    Sound level indicator is another unique project de-
        scribed   204
    Music synthesizers make use of ICs   206
    Many other interesting projects have been devised
        using ICs   210

SPECIAL BIBLIOGRAPHY OF HOBBY PROJECTS 212

GLOSSARY OF TERMS..214

BIBLIOGRAPHY 224

INDEX 230

DIGITAL IC's —

How They Work
and
How to Use Them

1

Understanding the Relationship Between Integrated Circuits and Transistors

In the early days of radio and electronics, circuits were simple, components were relatively large, including vacuum tubes, and no attention was paid to miniaturization.

There was very little standardization and each device or system was engineered from start to finish, stage by stage. Hand wiring became an expensive process. Engineers began to think of ways to standardize circuits to reduce engineering time and to search for easier wiring methods.

Printed circuits were invented and brought to a high degree of development. Standardizing circuits took a long time. First efforts involved the assembly of functional stages from discrete components in a compact rectangular package. These packages had plugs for plugging into a chassis socket. Encapsulation in epoxy or other plastic was used in some cases. These had some advantages in servicing in the field, saving in engineering time, but they never quite filled the bill.

Putting more than one transistor on a substrate started it all.

The advent of transistors emphasized the advantages of miniaturization. They also came at a time when electronic circuitry was rapidly increasing in complexity. Then someone, instead of making a two-element solid state diode or a three-element transistor, added another device on the same substrate—and the integrated circuit was born.

A plan view of an early IC is shown in Figure 1-1. This is a simple logic circuit employing two transistors, six diodes and two resistors. In a tiny package, it was a big step over discrete components.

FIGURE 1-1 (*Electronics* magazine, a McGraw Hill publication)

Actual and equivalent circuits aid in understanding ICs.

The capability of "growing" a very large number of transistors simultaneously on a small wafer of silicon is the key to the whole integrated circuit technology. Many of these transistors are used as transistors in the final circuits while others are connected to act as diodes, resistors and capacitors. In this way, all the elements for a wide range of R-C circuits are provided. However, a great many seemingly complex circuits are formed from combinations of standardized simple basic circuits. These simple basic circuits can be thought of as little black boxes performing a simple well-known function. The great majority of these functions are simply switching functions (digital).

Thus, we come back to the basic digital function—transistors used as switches. Typical IC packages contain anywhere from eight to several dozen transistors, not to mention additional transistors acting as diodes, resistors or capacitors. Devices not hitherto known are a by-product of IC technology. For example, transistors with two or more emitters can be produced, which have wide application as gates (switches).

In digital ICs the transistors are operated in one of two states: either completely cut off, or in full conduction (generally but not necessarily saturated). In these two states permissible ranges of input voltages required to produce these states are specified, as are the resulting output conditions. These input and output conditions depend on the way in which the transistors are used to accomplish the switching function.

The switching functions may be used in relatively simple and obvious circuits to gate signals in various ways or they may be combined to perform more complex functions. Digital computers are an example of how thousands of simple basic elements (switches or gates) can be combined to perform very complex functions.

While the digital IC, the switch or gate derived device or system, has many uses, there are several other types of IC. The analog or linear IC is also composed basically of transistors, but the transistors, instead of being operated on/off, are operated in their linear regions. Thus, operational amplifiers are linear ICs containing many transistors and resistors but designed for linear amplification of analog signals. Television and radio circuits are being built up using a few basic linear ICs such as IF amplifiers, audio amplifiers and so on. A development demonstrating the unique capabilities of ICs is the stereo multiplex decoder system, which does a superior job and at the same time eliminates tuning controls and all alignment problems. It is a combination of a number of sophisticated IC techniques, including frequency multiplication, high and low pass filtering and phase-locked loop operation.

Large Scale Integration (LSI) is an extension of the integrating techniques to ever more complex circuitry. Literally thousands of transistor and transistor synthesized circuits can be integrated on a tiny silicon chip. However, while there seems to be almost no limit to the number of transistors which can be grown on a chip, the law of averages imposes some practical limitations since an imperfect transistor can cause the whole device to be rejected. There is a balance between yield and other factors. Detection of imperfect transistors before final assembly permits rerouting circuits so that the chip will come out in acceptable form.

While one may think of LSI as useful where a large quantity of a complex circuit is required, such a circuit can also be economically produced in relatively small quantities from a predesigned computer program. Many such programs have been prepared and are in use. The microprocessor on a chip shown in Figure 1-2 is an example of advanced technology, employing several thousand transistors on a tiny chip.

A whole new technology was developed to manufacture ICs.

Since the most usual method of manufacture of integrated circuits uses the basic process of growing or depositing a large number of identical circuits on a single chip of silicon, this method will be described. It will serve to illustrate the advantageous techniques which go into their production.

First, a single crystal of pure silicon is grown to an appropriate diameter, generally between 1.5 and 2 inches in practical cases. This "ingot" is then sawed to produce a large number of identical round "wafers" of the order of 0.004 to 0.01 inch thick (1.5 to 2.0 inches in diameter). These wafers are highly polished to eliminate any scratches or saw marks, until a mirror-like surface is produced.

Second, the integrated circuits duplicated hundreds and even thousands of times are reproduced on each wafer, including lands for attaching leads.

FIGURE 1-2 (Motorola)

Third, tests are made to determine which circuits are good and the imperfect ones are marked or otherwise set for elimination in the final steps.

Fourth, the wafer is sawed into hundreds of dies or chips, each carrying a complete circuit.

Fifth, leads are attached and the individual ICs, either singly or in predetermined groups, are packaged by encapsulation in plastic, or sealed in a flat-pack or metal can.

Figure 1-3 compares a typical IC with the head of a common pin.

One of the well-known processes used for reproducing the integrated circuitry on the chip is known as *epitaxial*, which will now be described in some detail. In this process single-crystal silicon is grown on the wafer substrate by pyrolytic decomposition of a silicon halide. After forming this epitaxial layer, the steps typically include oxidation to form a thin protective surface; application of a photoresist; exposure through a mask pattern; removal of the photoresist in the delineated areas; etching to form windows for diffusion in the delineated areas of the surface oxide; removal of all photoresist and diffusion of predetermined impurities through the windows into the epitaxial layer; then start the cycle all over again, for as many as seven or eight times. The complete process takes from four to six weeks. A separate mask is required for each step: for example, steps including isolation diffusion, collector and resistor diffusion, emitter diffusion, contact window metalization, and interconnection-path metalization. Each mask must be aligned to within 0.00001 to 0.00002 inch or defective units will result. Diffusion impurities include such solids as boron trioxide and phosphorus pentoxide, such liquids as phosphorus oxychloride and boron tribromide, and such gasses as diborane and phosphene. Successful diffusion requires a precise impurity gradient over a precise depth. Temperature control during diffusion requires, say, 1200°C to within ¼ degree for several hours. Metalization for interconnections and external contact pads is generally provided by the deposition of aluminum.

Field effect ICs have some advantages.

It is interesting to note that the field effect transistor (originally called "fieldistor") and the bipolar transistor were conceived at about the same time. The original concept of the fieldistor paralleled the point-contact transistor concept except that the gate did not contact the semiconductor body acting purely electrostatically. Like the transistor, the fieldistor evolved into a junction type device, the JFET.

However, in a way the clock was turned back and the field effect transistor with a silicon dioxide insulator under the gate became again an electrostatic-operated device. The enhancement type MOS/FET goes right back to the original basic concept but is now implemented by critically controllable manufacturing techniques. Instead of an actual tunnel of semiconductor extending from the source to the drain, the source and drain are embedded in opposite conductivity

FIGURE 1-3
Typical IC Compared with the Head of a Common Pin

substrate—i.e., N material in P material or vice versa. The gate sits directly on top of the area (tunnel) between the source and drain and electrostatically induces a controlled current between these two points. The gate bias is a forward bias like a transistor, i.e., of the same polarity as the drain. With zero bias, like a transistor, there is little or no drain current. An increasing forward bias induces a proportionately increasing drain current across the diode junctions.

Another approach to the field effect transistor was pursued by W. Shockley of Bell Telephone Laboratories who backed up his work with some highly theoretical speculations. His initial results, reported in November 1952 (IRE), left considerable room for improvement. However, during the following year progress was made, and in August of 1953, G. C. Dacey and I. M. Ross described a satisfactory junction transistor with essentially the same basic characteristics as modern JFETs. These devices have found considerable use in analog devices such as in the input stage of operational amplifiers, as voltage controlled resistors, as constant current sources and as audio level control devices.

Defining small, medium and large scale integration.

The first integrated circuits were relatively simple devices embodying only a few devices—for example, a Darlington pair, a two-transistor differential amplifier or a number of diodes. However, when in the late 1960's the idea began to really catch on, more and more devices were integrated on a single chip and placed in a single package. There always was a balance between the number of perfect devices which could be successfully produced on a single wafer (the yield) and the final cost in the market place. Progress at times was slow but the ever increasing demand called for greater efforts. The number of devices per chip gradually increased from two to thousands.

The fact that integrated circuits cover such a broad range of complexity has led to classifying them in three rather broad categories: small scale integration (SSI), medium scale integration (MSI) and large scale integration (LSI). These classifications are rather loose and there may be substantial overlapping between them. To provide some basis for such designations the following definitions are offered:

SSI: Small numbers of gates or flip-flops up to an equivalent of, say, 12 or so gates.

MSI: Moderate number of flip-flops as in counters and registers up to (from SSI) an equivalent of, say, 100 or so gates.

LSI: Any IC having the equivalent of 100 gates or more.

Since LSI represents the highest order of development of the IC and is the subject of continuing development, further consideration will be given in Chapter 10 to this most intriguing subject.

Several basic packaging methods and forms were developed.

Early packaging of ICs followed closely the methods of packaging transistors being used at the time. While the transistor had three leads, the IC had more. The technique was to provide a header with the required number of leads arranged in a circle; after placing the chip on the header and connecting the leads to the pads on the IC chip, the package was completed with a round metal can. These metal can types generally had 8, 10 or 12 pins requiring compatible sockets. These metal can ICs had the disadvantages that insertion into the sockets were not foolproof and extension beyond 12 pins seemed impractical.

The next step was to mount the IC chip in a flat-pack. The flat-pack with 14 or 16 terminals provided greater flexibility in circuit design and permitted more complex circuitry in a given package. The flat-pack was generally constructed as a ceramic sandwich providing excellent hermetic sealing and long life protection. The rectangular form provided straight line and hence improved lay-out of the printed circuit boards. It was not a readily socketable device and was soldered or welded to the printed circuit board connections. This made field servicing difficult. ICs do break down in use and must be replaced. One of the easiest ways to find circuit trouble is to take out a suspected package and substitute one known to be good. The flat-pack mitigated against this procedure.

The next step was the dual-in-line or DIP package. This is not only easily socketed but also, since it is rectangular, has all of the advantages in printed circuit board wiring of the flat-pack. Standard DIP's include 14 pin, 16 pin, 8 pin (half of 16), 24, 40 and other. There seems to be no particular limit. Sockets can be extended merely by adding end to end. The final limit, if there is to be one, would seem to be how much does it pay to put in one package. DIP packages are made in hermetically sealed ceramic form, but much more popularly in plastic. The DIP socket is relatively simple and effective. However, with larger and larger numbers of pins the pressure required to press the DIP into its socket increases proportionately. One answer to an increasing difficulty in this regard is the so-called zero-force sockets in which the socket-gripping contact pressures are released for easy insertion of the package. Once the package is in place, socket spring pressure is brought to bear on the package pins by screwing the cover down.

2

Using Logic to Explain the Operation of ICs

Logic broadly interpreted means the science of reaching reasoned conclusions by correctly thinking through a set of postulates. However, the logic we are dealing with can be more correctly termed "pure logic" since with a given set of conditions the resultant is assured. In other words, there are no unanswered questions; the answer is foregone. For example, we know that if we place two switches in series with a battery and a lamp and close *both* of these switches (see Figure 2-1), the lamp will light. This is AND logic: when both switches (A and B) are closed the lamp (C) lights. By convention, an equation for this is written A · B = C. This is a basic equation of Boolean Algebra, a system of logical notation first proposed by G.

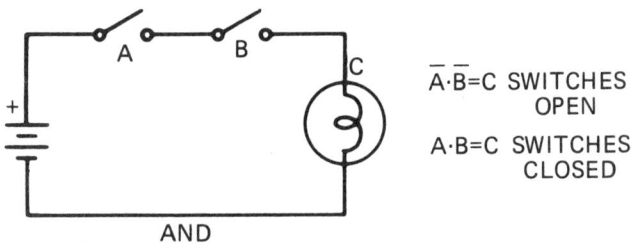

$\overline{A} \cdot \overline{B} = C$ SWITCHES OPEN

A·B=C SWITCHES CLOSED

AND

FIGURE 2-1

Boole in 1854 and adopted by the integrated circuit technology. Note that the AND function is shown as A · B (other notations are in use,

28

such as simply AB). If we place the switches in parallel (see Figure 2-2), closing either switch will light the lamp. This is called the OR function and the equation is written A + B = C (here OR is represented by the + sign). The third function is inversion or NOT written $\overline{A}$, meaning A is inverted (in the case of the switches, $\overline{A}$ means the switch is open). Since the collector potential is the reverse of the base potential in a transistor, inversion is usually obtained as shown in Figure 2-3.

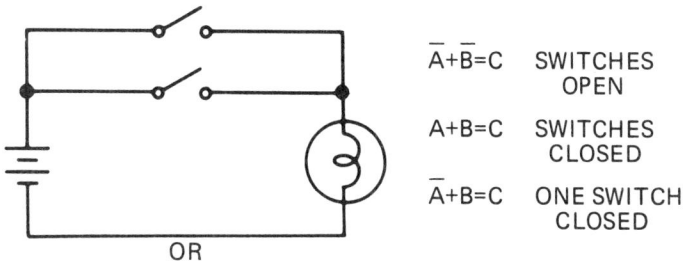

$\overline{A}+\overline{B}=C$	SWITCHES OPEN
$A+B=C$	SWITCHES CLOSED
$\overline{A}+B=C$	ONE SWITCH CLOSED

OR

FIGURE 2-2

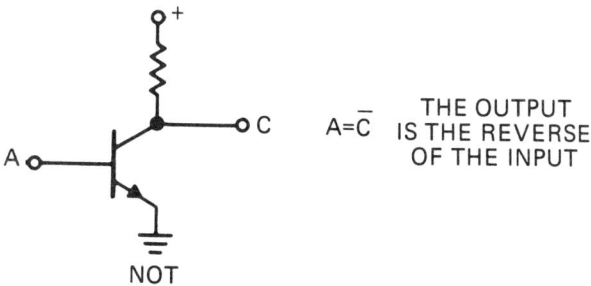

$A=\overline{C}$ THE OUTPUT IS THE REVERSE OF THE INPUT

NOT

FIGURE 2-3

Symbols adopted for greater usefulness.

Integrated circuit technology uses symbols earlier devised but particularly refined for use in this new art. As often happens in a similar situation, i.e., a new and developing art using a language of symbols, a wide variety of symbols have been used by different engineers and companies. There are four symbols which are used over

and over again. These are symbols for AND gate, OR gate, INVERT
and DELAY. The most widely used standard symbols are those
adopted as MIL-STD-806B and they will be used in this book (Figure
2-4), except that a simple circle (instead of triangle and circle) will be
used for INVERT. The triangle is the conventional amplifier symbol.

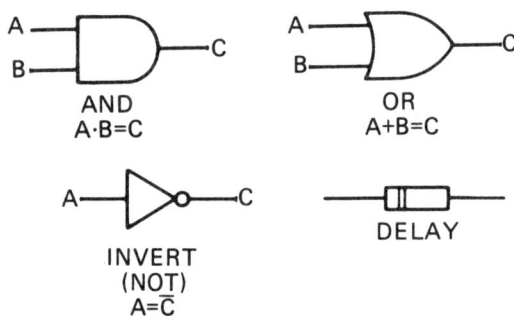

FIGURE 2-4

The invert symbol as a circle can be added to the AND and OR
symbols to form NAND (AND inverted) and NOR (OR inverted) as
shown in Figure 2-5.

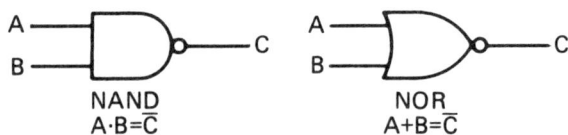

FIGURE 2-5

Symbolic representation of the myriad of complex ICs is made
simply by a rectangle carrying the name or abbreviation for the name
of the circuit. Manufacturers generally complicate matters by merely
showing their part number on the product, leaving you to figure out
just what the rectangle represents.

How simple ICs provide the basic logic gates.

The three basic elements of IC logic are AND gates, OR gates and INVERTERS, shown symbolically in Figure 2-4. While other symbols have been and are used, these are most widely used and are readily recognizable in any configuration. They will be used throughout this book.

When an inverter is added to the output, the AND gate becomes a NAND gate (AND and INVERT) and the OR gate becomes a NOR gate (OR and INVERT), as shown symbolically in Figure 2-5.

As may already be apparent, logic can exist in only one of two possible states—OFF or ON. Thus, substituting 1 for A, 1 for B, 1 for C, 0 for $\bar{A}$, 0 for $\bar{B}$ and 0 for $\bar{C}$, we can write the equations again like this: $1 \cdot 1 = 1$; $1 \cdot 0 = 0$; $1 + 1 = 1$ and $1 + 0 = 1$. This is called "positive" logic. Negative logic is the inverse and will be discussed later.

Truth tables are one practical aid in analyzing circuits.

Another device which permits more complete understanding of logic performance is the "truth table." This is a formal statement of the relative conditions of all inputs and outputs of a given gate. It is a powerful tool, especially when dealing with complex systems of gates.

Figure 2-6 is the truth table for the AND gate. There are four possible combinations of input logic, and each combination produces a predictable output. When A and B are both 0 (off or open), C is 0; when A is zero and B is 1 (closed), C is still zero (from Figure 2-1 it can be seen that if either switch is open, there is no output); when A is 1 and B is 0, C is still 0; but when A is 1 and B is 1, C is also 1. To

$A \cdot B = C$ (AND)
$A \cdot B = \bar{C}$ (NAND)

A	B	C	$\bar{C}$
0	0	0	1
0	1	0	1
1	0	0	1
1	1	1	0
AND			NAND

FIGURE 2-6

recapitulate, in an AND gate *all* inputs must be 1 to produce an output 1. Inverting the output of an AND gate produces an inversion of C ($\overline{C}$) as shown in the NAND column.

The truth table for the OR gate is shown in Figure 2-7. Here C is 0 if both A and B are 0, but C is 1 if either or both A and B are 1. Again inverting C to $\overline{C}$ produces a NOR output (OR inverted in the output) as shown in the NOR column.

A+B=C (OR)
A+B=$\overline{C}$ (NOR)

A	B	C	$\overline{C}$
0	0	0	1
0	1	1	0
1	0	1	0
1	1	1	0
	OR		NOR

FIGURE 2-7

These gates may be implemented in many ways by discrete component or integrated circuits. In positive logic 1 means high or ON while 0 means low or OFF. In practical circuits the ON or high (1) voltage might mean anything between + 3.6 and + 5.0 volts and the OFF or low (0) voltage might be anything below + 1.8 volts down to zero. In order to provide the unambiguous two states it is not necessary to have exact voltages. This is one of the great advantages of digital logic; it is ON or OFF, high or low, and no in-between states need exist or be recognized.

An actual circuit will help to visualize the gate concept. Figure 2-8 shows a diode-resistor (DRL) logic AND gate in the four states expressed in the truth table of Figure 2-6. When both cathodes of diodes A and B are grounded (a) they conduct, lowering the potential to C to 0 (actually about 0.6 volt, the forward conduction voltage of a silicon diode, but logic 0). Likewise as shown at b and c, if either cathode is grounded, conduction lowers C to logic 0. However, if both cathodes are connected to + (high or logic 1), neither diode conducts and C is high or logic 1. This is one of the most elementary and basic of the possible AND gates. It should be noted that A and B are connected either to ground (0) or + (1). A basic requirement in most logic gate circuitry is that gates are dynamically placed at 0 or 1, although in some cases an open gate can be assumed to be at logic 1. Another requirement generally is that when a gate is forced to 0 the circuit doing the forcing must sink some current, the actual amount depending on the type of logic.

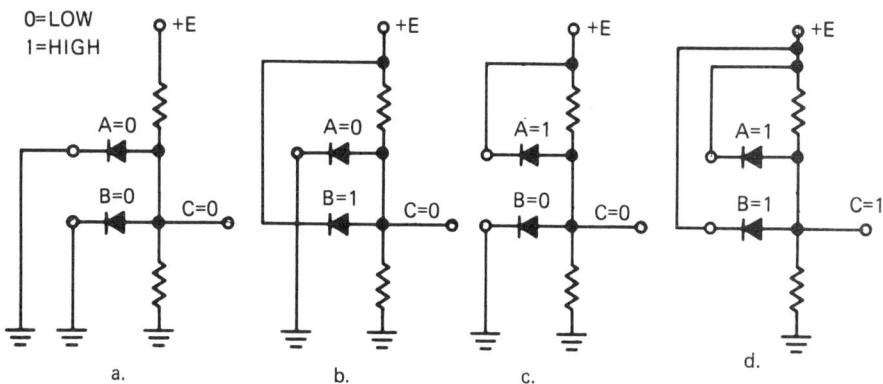

DIODE AND GATE   A·B=C
(POSITIVE LOGIC)

FIGURE 2-8

Reversing the diodes and omitting the + E source as in Figure 2-9 provides a simple diode OR gate in accordance with the truth table of Figure 2-7. If both A and B are zero, C is zero. The applicable equation is A + B = 0. If either or both A or B are 1, C is 1.

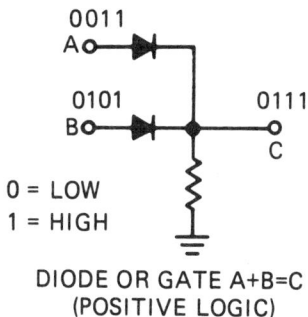

DIODE OR GATE A+B=C
(POSITIVE LOGIC)

FIGURE 2-9

The numbers after A, B and C in Figure 2-9* show the corresponding input and output conditions in order; the first numbers are 0
for A, 0 for B and 0 for C; the second are 0 for A, 1 for B and 1 for C;
and so on. This notation is convenient and visual and aids substantially
in following signal conditions, particularly in circuits involving combinations of gates.

The diode OR gate is a true OR gate for positive logic, i.e., the
output is controlled by the more positive of the input signals. While
this and other gates are being shown as two-input gates, more inputs
can obviously be added making three, four or more input gates where
the application calls for them. One advantage of this simple diode gate
is that it is not critical with respect to input voltage as long as the logic
1 exceeds the diode threshold voltage and the output can tolerate this
input voltage. One disadvantage is that the input must drive the output
directly with no gain-producing elements. For flexibility of design and
combination, transistors are widely used in gate circuits, and most IC
logic includes transistors.

Figure 2-10 shows a simple transistor OR gate; a high voltage (1)
on either A or B input or both will produce an output. In order to
prevent excessive loading on the drive circuits, resistors are placed in
series with the bases. This is known as resistor-transistor logic or RTL.

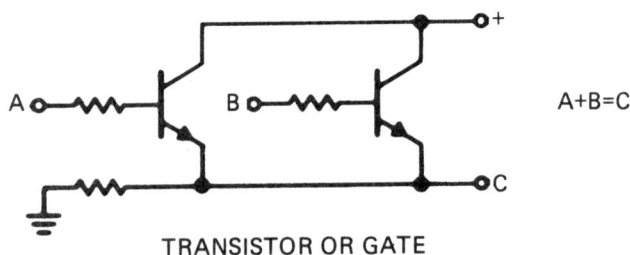

TRANSISTOR OR GATE

FIGURE 2-10

Since the resistors in series with the transistor base capacitance slow
down the response of the circuit to fast pulses, capacitors are sometimes used shunting the resistors (RCTL).

---

*Presenting a new way to visualize logic levels in interconnected circuits.

Figure 2-11 shows a simple transistor AND gate also for positive logic. These last two gate circuits employ active elements in the form of transistors. The output is isolated from the inputs. Input loading is greatly decreased for a given load impedance. By far the greater part of IC gate technology has evolved from the transistor gate concept.

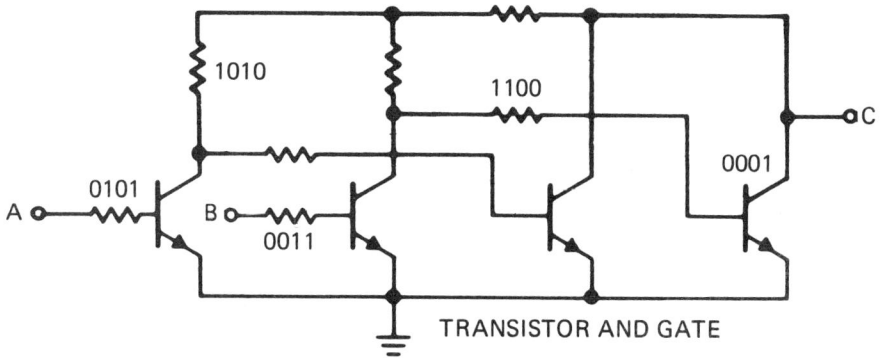

TRANSISTOR AND GATE

FIGURE 2-11

The third basic logic element is the inverter or NOT gate. This gate does not change the content or combination of logic applied but merely inverts it. Figure 2-12 shows a simple transistor inverter or NOT gate. Input applied to the input appears inverted at the output. The truth table shows that if the input is 0, the output is 1, and if the input is 1, the output is 0.

A	$\overline{A}$
0	1
1	0

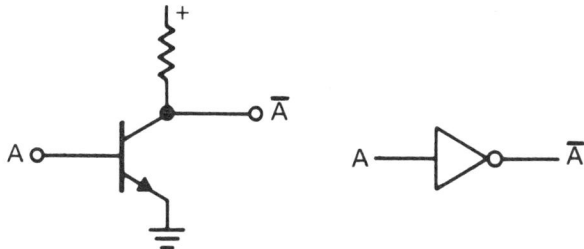

TRANSISTOR NOT GATE (INVERTER)

FIGURE 2-12

Why inverting gates require fewer transistors than non-inverting gates.

Logic using transistors has several advantages over diode or other types of logic. Transistor logic provides high input impedance, low output impedance and current gain facilitating combinations of several or many gates. But a transistor gate is generally used as an inverting gate—that is, after the OR or AND function has been performed, the output is inverted. Such gates are known as NOR and NAND gates and their equations are written as $C = \overline{A + B}$ and $C = \overline{A \cdot B}$ respectively. They would not be written $C = \overline{A} + \overline{B}$ or $C = \overline{A} \cdot \overline{B}$ since the bar over A and B independently means the inversion is performed on A and B independently (at the input) rather than after being combined.

Several types of logic have evolved with the growing capabilities of IC production. One of these is the so-called TTL or transistor/transistor logic which makes use of the peculiarly IC adaptable technique of growing multiple emitters on a transistor. Figure 2-13 shows the basic equivalent circuit of a TTL NAND gate. The truth

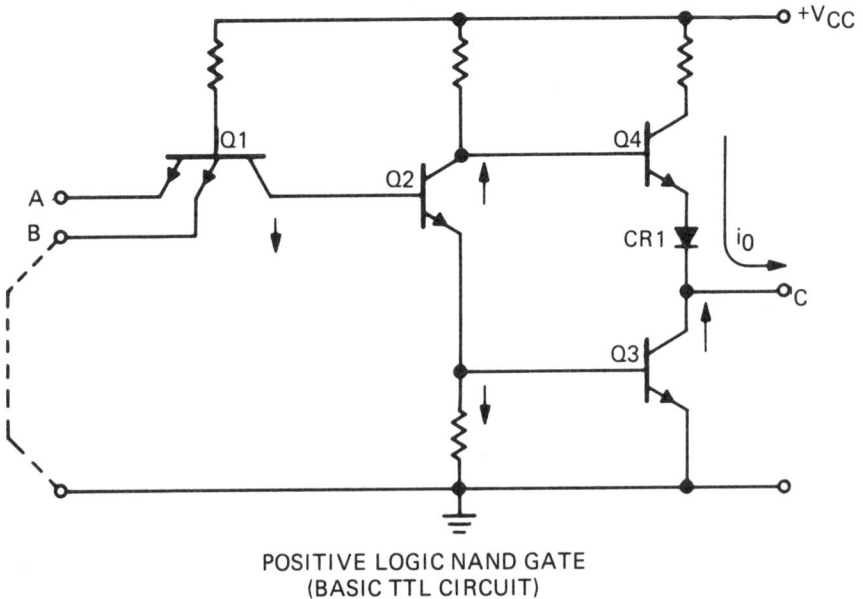

POSITIVE LOGIC NAND GATE
(BASIC TTL CIRCUIT)

FIGURE 2-13

table for this circuit is like Figure 2-6 but with a fourth column labeled $\overline{C}$, which is in each case the inverse of the Cs in the third column.

Looking at Figure 2-13, if either or both emitters A or B are grounded (0), current will flow through the emitter resistor of Q_1 and its collector will go low. Transistor Q_2 thus deprived of base drive will not conduct so that its emitter will go low, turning off Q_3, and its collector will go high, turning on Q_4. With Q_3 off and Q_4 on, full current (logic 1) will flow to terminal C. Only if both emitters (all emitters) of Q_1 are high (1) will the conductions and potential levels be reversed, producing output zero (0). Diode CR_1 is provided to insure an operating voltage drop for transistor Q_2. This NAND gate is a common and useful IC gate and is readily produced by IC technology with up to five emitters (inputs).

NAND gates can be used in many ways in logic systems. For example, the logic equations, charts (to be explained later) or truth tables can be implemented direclty; negative logic may be used; or the negated gates (NAND and NOR) can be converted back to positive logic gates (AND and OR) by means of inverters.

The NAND or negated AND symbol, equation and truth table are shown in Figure 2-14. The NOR or negated OR symbol, equation and truth table are shown in Figure 2-15.

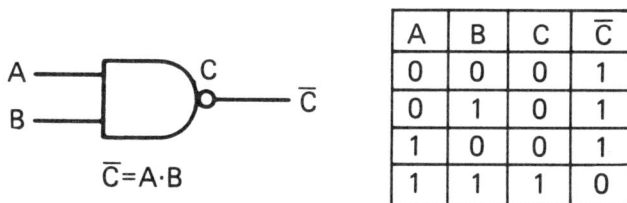

A	B	C	$\overline{C}$
0	0	0	1
0	1	0	1
1	0	0	1
1	1	1	0

$\overline{C}=A\cdot B$

FIGURE 2-14

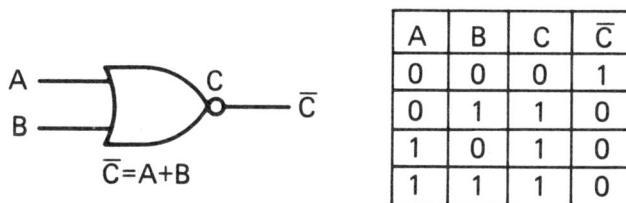

A	B	C	$\overline{C}$
0	0	0	1
0	1	1	0
1	0	1	0
1	1	1	0

$\overline{C}=A+B$

FIGURE 2-15

Negative logic helps in understanding inverting ICs.

By convention, positive logic means that a 1 is high and an 0 is low and follows naturally from the mathematical notation 1 and 0. However, negative logic is also used and has some interesting implications. In negative logic 1 is low and 0 is high. The interesting and significant result is that an AND gate for positive logic becomes an NOR gate for negative logic and an OR gate for positive logic becomes an NAND gate for negative logic. Also NAND becomes OR and NOR becomes AND. So when it is stated that a given gate is, say, NAND "for positive logic," it will also be OR "for negative logic."

Figure 2-16 shows the AND gate for positive logic of Figure 2-8 with negative logic applied. It will be seen that with negative logic it is an OR gate. Similarly, Figure 2-17 shows how the positive logic OR gate of Figure 2-9 becomes a negative logic AND gate.

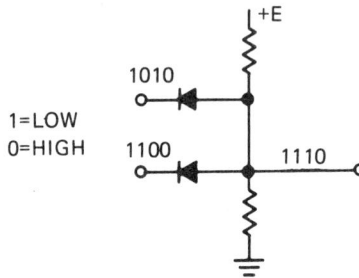

NEGATIVE LOGIC OR GATE

FIGURE 2-16

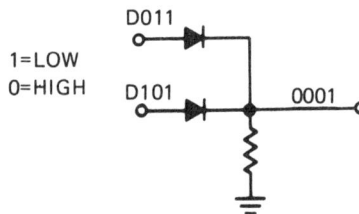

NEGATIVE LOGIC AND GATE

FIGURE 2-17

Figure 2-18 shows two truth tables for the AND/NAND gate of Figure 2-14. The positive logic table is that of an AND or NAND gate and the negative logic table for the same gate is that of an OR or NOR gate. The reverse is also true, i.e., a positive logic OR/NOR gate is equally a negative logic AND/NAND gate (see Figure 2-15).

A	B	C	$\bar{C}$
0	0	0	1
0	1	0	1
1	0	0	1
1	1	1	0

AND/NAND
POSITIVE LOGIC

A	B	C	$\bar{C}$
0	0	0	1
0	1	1	0
1	0	1	0
1	1	1	0

OR/NOR
NEGATIVE LOGIC

FIGURE 2-18

The above explains why it is not necessary to have both AND and OR or NAND and NOR gates to provide both functions. The TTL NAND gate of Figure 2-13 is a gate particularly adapted to IC technology. The fact that it can also be used as an OR gate simply by inverting the input signals (making negative logic signals) makes it of quite general usefulness as a gate. More will be said on this subject in Chapter 3, where various types of logic are described and compared.

The exclusive OR gate is a very useful device.

A gate which is actually composed of a combination of simple gates is the exclusive OR. With two inputs this gate produces an output if either of the two inputs is 1 but not if both are 0 or 1. There are many ways in which this gate can be implemented from AND, OR gates and inverters. One which uses the four NAND gates of a 7400 DIP package is shown in Figure 2-19. The relative signal levels on the various leads are shown as well as the truth table, equation and symbol. With more than two inputs, the exclusive OR gate can be provided by combining two input exclusive OR gates as shown symbolically in Figure 2-20 for three inputs or four inputs. Further expansion is quite obvious.

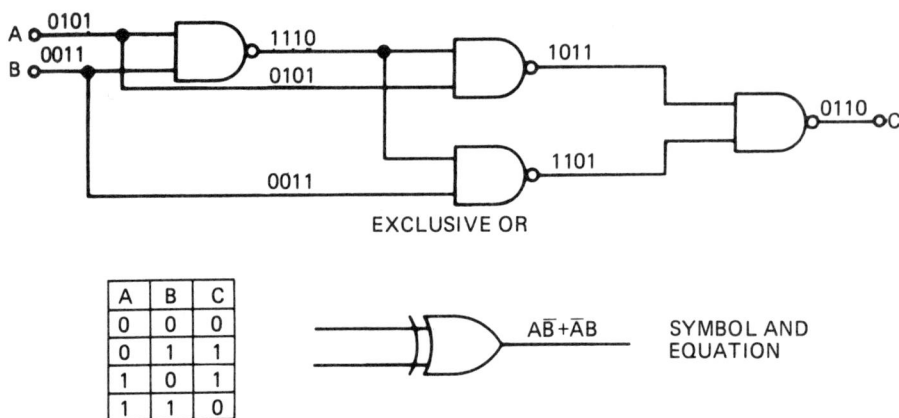

EXCLUSIVE OR

A	B	C
0	0	0
0	1	1
1	0	1
1	1	0

$A\overline{B}+\overline{A}B$ SYMBOL AND
EQUATION

FIGURE 2-19

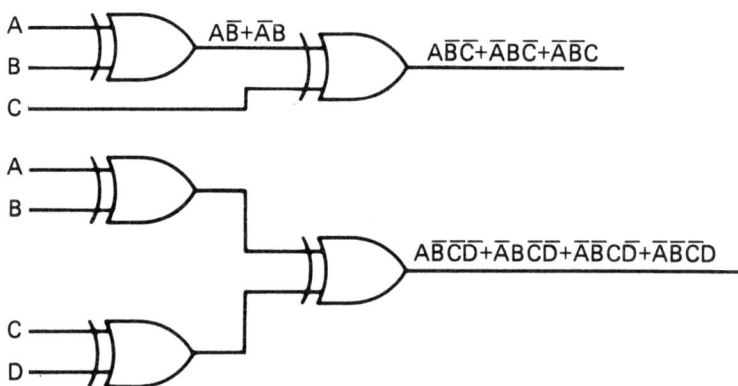

$A\overline{B}+\overline{A}B$

$AB\overline{C}+\overline{A}B\overline{C}+\overline{A}BC$

$AB\overline{C}\overline{D}+\overline{A}B\overline{C}D+\overline{A}BC\overline{D}+\overline{A}\overline{B}CD$

FIGURE 2-20

Eight permutations of basic gates with equations and equivalences.

Eight permutations of the two basic gates and inverter are shown in Figure 2-21, together with the applicable equations and truth tables. In the first line are shown the AND, OR, NAND and NOR two input gates with symbols, equations and truth tables. These gates are shown for positive logic inputs. Immediately below each gate is shown a gate which is equivalent in that it has the same truth table as the one above.

EIGHT PERMUTATIONS OF BASIC GATES

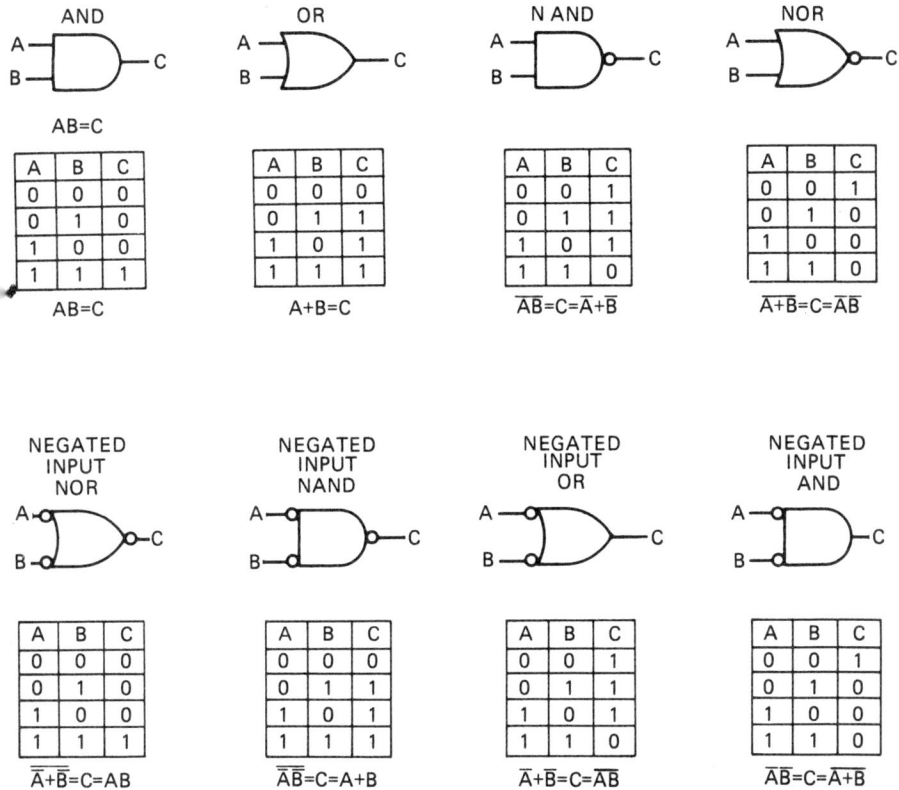

AND — AB=C

A	B	C
0	0	0
0	1	0
1	0	0
1	1	1

AB=C

OR

A	B	C
0	0	0
0	1	1
1	0	1
1	1	1

A+B=C

N AND

A	B	C
0	0	1
0	1	1
1	0	1
1	1	0

$\overline{AB}=C=\overline{A}+\overline{B}$

NOR

A	B	C
0	0	1
0	1	0
1	0	0
1	1	0

$\overline{A+B}=C=\overline{A}\,\overline{B}$

NEGATED INPUT NOR

A	B	C
0	0	0
0	1	0
1	0	0
1	1	1

$\overline{\overline{A}+\overline{B}}=C=AB$

NEGATED INPUT NAND

A	B	C
0	0	0
0	1	1
1	0	1
1	1	1

$\overline{\overline{A}\,\overline{B}}=C=A+B$

NEGATED INPUT OR

A	B	C
0	0	1
0	1	1
1	0	1
1	1	0

$\overline{A}+\overline{B}=C=\overline{AB}$

NEGATED INPUT AND

A	B	C
0	0	1
0	1	0
1	0	0
1	1	0

$\overline{A}\,\overline{B}=C=\overline{A+B}$

FIGURE 2-21

In each of these four gates in the lower row the inputs are inverted so that what the gate sees in each case is equivalent to negative logic. It will be seen that the negated input NOR gate is functionally equivalent to an AND gate (also this shows that a NOR gate is equivalent to an AND gate for negative logic); the negated input NAND gate is equivalent to an OR gate; the negated input OR gate is equivalent to a NAND gate; and the negated input AND is equivalent to a NOR gate. The equations for the various gates also show these equivalencies. Incidentally, they also prove two of DeMorgan's Laws, namely $\overline{AB} = \overline{A} + \overline{B}$ and $\overline{A}\,\overline{B} = \overline{A + B}$.

3
Understanding How Circuits Have Been Used in Logical ICs

Logic ICs have evolved from earlier vacuum tube circuits.

There has ben a substantial amount of evolution in the IC digital logic module field since the early beginnings in 1960. The gates which were formerly switches, vacuum tubes or diodes have largely become transistor gates, either bipolar or unipolar. Varying needs and circuit philosophies have resulted in a number of distinct types of logic circuit and circuit equivalents. Some types are popular for a time but have been superseded. Others have stood the test of time. It is likely that there will be further evolution and it is just as likely that no one type will be the final answer to all needs.

It would be impossible to completely cover and evaluate all of the logic forms which have been developed and are now in use. It is important to point out, however, that while there are many, understanding some of the major types will assist in understanding still others and will point the way to knowledgeable use of the available types.

Important factors affecting the types of logic packages being manufactured include circuit density, power requirements, package dissipation, fan-in and fan-out capabilities, compatibility with other types, noise generation, noise immunity, speed of response, cost and availabiility.

It will become increasingly apparent as various logic types are presented and analyzed that they are all derived from basically simple transistor circuits. One important reason that they can be simple is the fact that digital circuits operate in one of two possible output states. The linearity and distortion criteria which must be met in analog circuits just don't exist. All logic IC gates except the diode gates use a transistor in the output, and this transistor is merely turned on or off. In many cases the ON condition is saturation, making operation still less exacting. The beauty of the IC concept lies in creating complex structures from many simple basic elements.

The concept of logic and its implementation by means of diode and transistor circuits have been set forth in Chapter 2. Now we will get down to cases by describing the various types of logic circuits which have been devised to carry out the basic concepts of logic.

How transistors are used to provide the three major building blocks.

The three basic building blocks are inverter, OR gate and AND gate. From these three, all other logic blocks can be formed, as, for example, counters, registers, and so on. It turns out that gates for most purposes use transistors, using their gain to provide sharp transitions, reconstitution of the signal and so on. However, a single transistor stage used in a voltage gain mode also inverts the phase of the signal. This means, for example, instead of an OR gate we get a NOR gate (OR with inverted output). While adding another transistor amplifier stage will convert the NOR gate to an OR gate, the cost of the gate has been increased. Thus, if we can live with NOR gates and NAND gates (AND with inverted output), we can save very substantially in the cost of logic gates. This explains why most available gates are NOR or NAND gates. We have learned to live with them.

Many logic types have been developed, including RTL-RCTL-DTL-TTL-ECL-HTL-UTL-MOS/FET-C/MOS.

RTL. Resistor-transistor logic is a family of logic composed of basic and simple resistance coupled transistor circuits. The basic inverter as shown in Figure 3-1 is a transistor with a collector resistor as an output load and a base resistor to limit input current. When input A is grounded (0), the transistor passes little or no collector current and output B is high (1), essentially equal to VCC. When A is high (1)

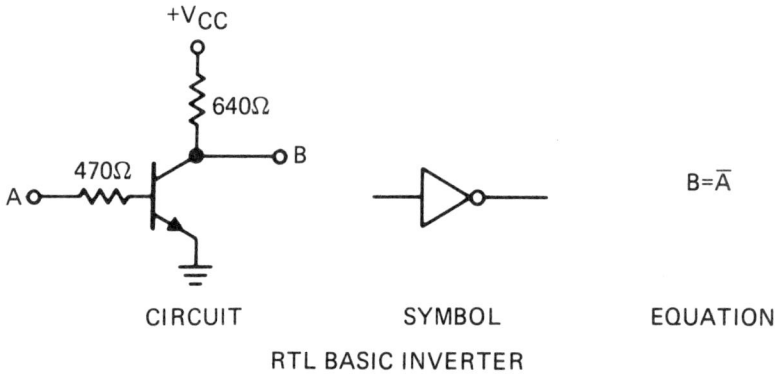

CIRCUIT   SYMBOL   EQUATION

RTL BASIC INVERTER

FIGURE 3-1

(VBE < 1 < VCC), the transistor passes current, generally substantially saturation, and B goes low (0) (B = VBE). VBE is the base to emitter voltage during conduction (saturation).

The basic RTL gate uses two or more transistors with collectors in parallel and separate resistance input base circuits as shown in Figure 3-2. The truth table for this gate is also shown along with the equation.

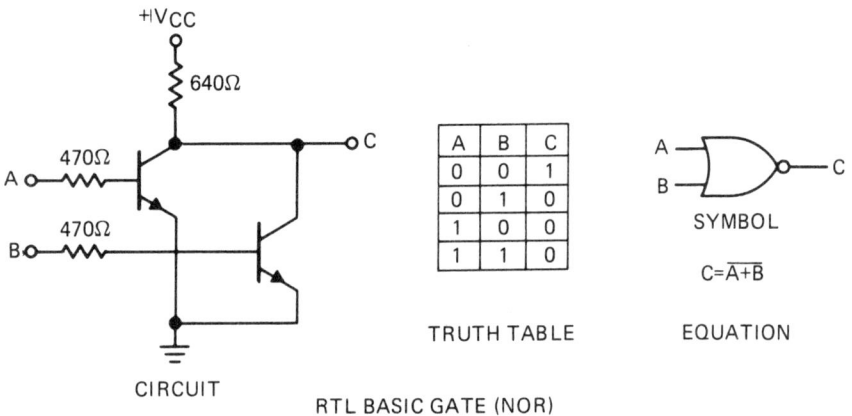

A	B	C
0	0	1
0	1	0
1	0	0
1	1	0

CIRCUIT   TRUTH TABLE   EQUATION

$C=\overline{A+B}$

RTL BASIC GATE (NOR)

FIGURE 3-2

If A and B are low (0), the output is high. For all other conditions, the output is low. This will be seen to be a NOR gate for positive logic and therefore an AND gate for negative logic.

Increased output for buffering and/or inverting is provided by a three-transistor gate as shown in Figure 3-3. By using two active transistors in the output circuit in place of the usual single transistor and resistor, a push-push circuit is created with greater output current capability. When input A is low (0), transistors Q_1 and Q_2 are non-conducting (high impedance) and Q_3 is conducting. Output B is then high (1), having a low impedance to $+V_{CC}$ and a high impedance to ground. When A is high (1), Q_1 conducts remove the base bias from Q_3, making it high impedance and causing Q_2 to conduct, bringing output B low (V_{BE}). Substantially the only current drawn in the output circuit will thus be the external load connected to B and it will be supplied through the 100 ohm collector resistor and Q_3 in series in its conducting state.

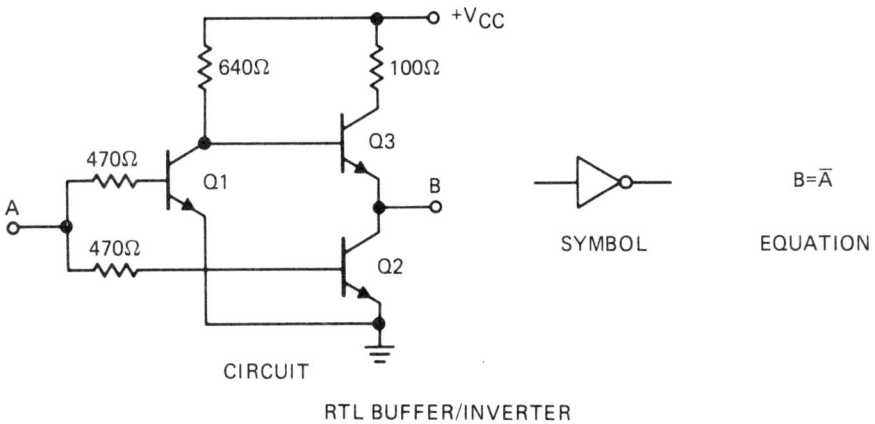

RTL BUFFER/INVERTER

FIGURE 3-3

RTL logic includes, in addition to the basic inverters and NOR gates, a wide range of flip-flops, encoders, decoders, counters, dividers, shift registers and so on. As stated above, the basic circuit unit is a simple transistor amplifier and the complex circuits are formed by interconnecting many of these simple basic circuits. Input and output considerations are those of the simple circuit. Internal operations are the interplay of these logic building blocks. Circuit details are generally not important except for those of input and output.

RTL logic uses an input base resistor to limit input current. A commonly used V_{CC} is $+3.6$ volts. The base resistor and hence the input (load) current depends on the application. Low power applications use higher values of base resistor and hence draw less current. They likewise use higher values of collector resistance and therefore can deliver less output current. Where more loads are to be supplied than can be handled by the single output transistor of the gate, buffers may be used. It must be remembered, however, that the buffer may also be an inverter.

RCTL. The relatively slow response of the RTL, due to the time constant of the base resistor and base diode capacitance, may be overcome to a certain degree by placing a small capacitor across the base input resistor. This form of logic circuit is accordingly known as RCTL. This has not proved to be a popular IC form due to the difficulties inherent in integrating capacitors. The ECL and TTL logic forms provide high speed and are more compatible with IC integrating techniques.

DTL. One of the first digital logic circuits to be integrated was the diode-transistor logic circuit (DTL) which comprised mainly the use of a transistor inverting stage after a diode gate. Figures 3-4 and 3-5 show circuits and truth tables for two DTL gates, one positive logic NAND and the other positive logic NOR. This is a relatively simple gate, capable of supporting a large number of diode inputs (large fan-in capability), and a reasonable number of devices can be driven from the transistor output (fan-out). The lower the value of the collector resistor, the more output devices can be driven but also the larger the "ON" dissipation. This is a relatively low speed device (when compared to devices designed to maximize speed) due to diode storage capacitance and the R-C time constant of the base capacitance and series resistor. Bypassing this resistor increases the response speed.

TTL. One of the major developments which resulted from integrated circuit capabilities is the transistor-transistor logic circuit (TTL). The technique of producing multiple emitters was developed as a part of the art of IC fabrication. The discovery that several emitters could be grown on a single transistor base led to the TTL structure. A typical NAND gate of this type is shown in Figure 3-6. Specifications sometimes are written NAND/OR, meaning NAND for positive logic and OR for negative logic. The TTL logic employing multiple emit-

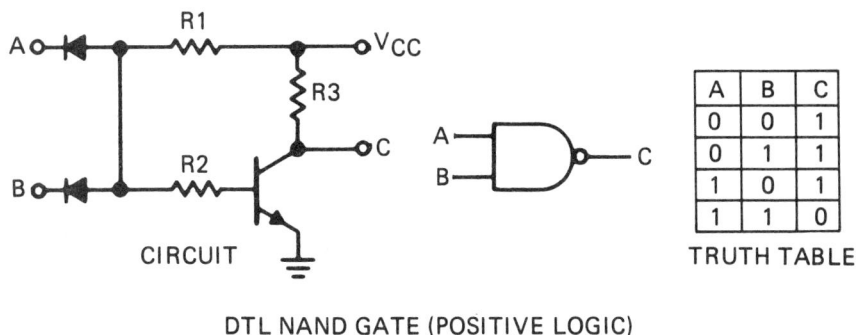

A	B	C
0	0	1
0	1	1
1	0	1
1	1	0

TRUTH TABLE

CIRCUIT

DTL NAND GATE (POSITIVE LOGIC)

FIGURE 3-4

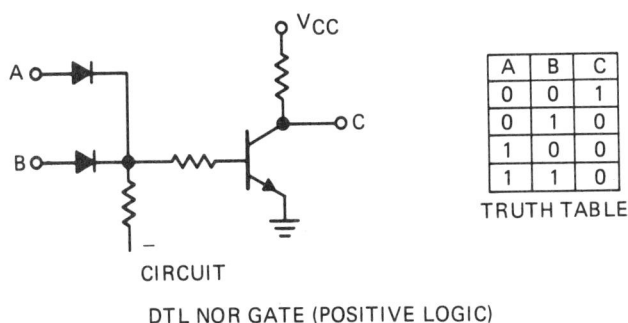

A	B	C
0	0	1
0	1	0
1	0	0
1	1	0

TRUTH TABLE

CIRCUIT

DTL NOR GATE (POSITIVE LOGIC)

FIGURE 3-5

ters (as many as eight providing eight inputs) is generally a NAND gate for positive logic. A modified form is provided to make an AND-OR-INVERT gate (Figure 3-7) which is expandable (provision for more inputs). A suitable expander looks and functions like the first two transistors of the NAND gate, as shown in Figure 3-8. The TTL family is considered to be characterized by relatively high speed, high noise margin and low power dissipation.

TTL gates function in a manner quite similar to the DTL gate. The

emitters of the input transistor of TTL logic play the part of the input
diodes of DTL logic. Referring again to Figure 3-4, if either input A or
B—or both—are grounded (0), the base end of resistor R_1 is also
grounded, and no bias is supplied to the transistor, no collector current
is drawn and the output C is at the potential of V_{CC} (1). However, if
both A and B are high (1), the base of the transistor receives an on bias
through R_1 causing collector current to flow through R_3. This causes a
voltage drop in R_3 and output C drops to the V_{CE} saturation voltage of
the transistor or logic 0.

Following the potentials through the TTL gate of Figure 3-6 is
very similar. If either or both input emitters A and B are at ground
potential (0), both the base and collector of transistor Q_1 will be low
(0) and a low bias will be placed on follower transistor Q_2. Since Q_2
will not draw current under this condition, its collector will be high (1)
and its emitter will be low (0). Thus, Q_3 will have a high base bias
turning it on and Q_4 will have a low base bias turning it off. With Q_3
conducting and Q_4 high impedance, current will flow from V_{CC} to
output C, making it high (1) and an inversion of the inputs. If both
inputs are high, the conditions are reversed and C is low (0). This is a
NAND situation as shown in the truth table. The numbers 0 and 1
shown adjacent to the various transistor elements show in order the
simultaneous conditions of 0 and 1 as described above.

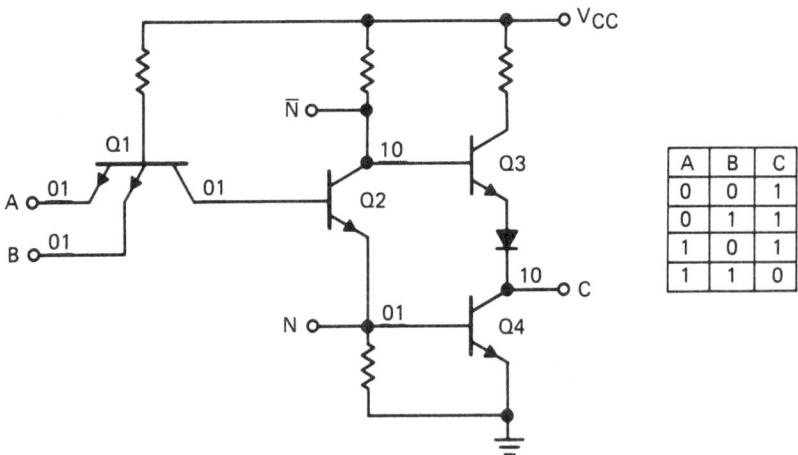

A	B	C
0	0	1
0	1	1
1	0	1
1	1	0

TTL NAND GATE (POSITIVE LOGIC)

FIGURE 3-6

Derived from the TTL NAND gate just described (Figure 3-6) is the so-called AND-OR-INVERT gate as shown in Figure 3-7. The output R remains at 1 unless A and B are both 1 or C and D are both 1. The equation for this gate is then R = $\overline{(AB) + (CD)}$.

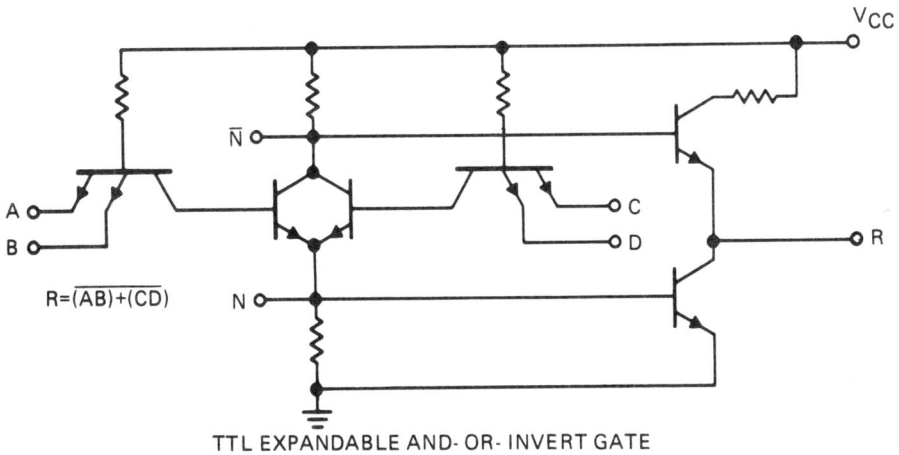

TTL EXPANDABLE AND- OR- INVERT GATE

FIGURE 3-7

Both the NAND gate of Figure 3-6 and the AND-OR-Invert gate of Figure 3-7 are what is known as "expandable." That is, additional inputs may be provided by means of a simple two-transistor expander as shown in Figure 3-8. When the $N\overline{N}$ terminals are connected to the corresponding $N\overline{N}$ terminals of the NAND gate of Figure 3-6 or the AND-OR-Invert gate of Figure 3-7, three more inputs (ABC) are provided OR'ed with the original inputs. The equations with the expander become C = $\overline{(AB) + (ABC)}$ and R = $\overline{(AB) + (CD) + (ABC)}$ respectively.

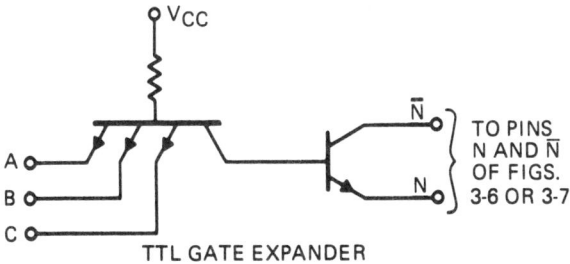

TTL GATE EXPANDER

FIGURE 3-8

ECL/MECL. The most interesting aspect of emitter-coupled logic (ECL) is its high speed capability. The important trade-off is power requirements and package dissipation which run comparatively high. The circuit is based on a differential pair transistor input and emitter follower outputs. Thus, inverting and non-inverting inputs are provided together with non-inverted (OR) and inverted (NOR) output transistors. The basic circuit is shown in Figure 3-9. The negative operating bias and negative input signals are not compatible with most logic which operates on the positive side of ground.

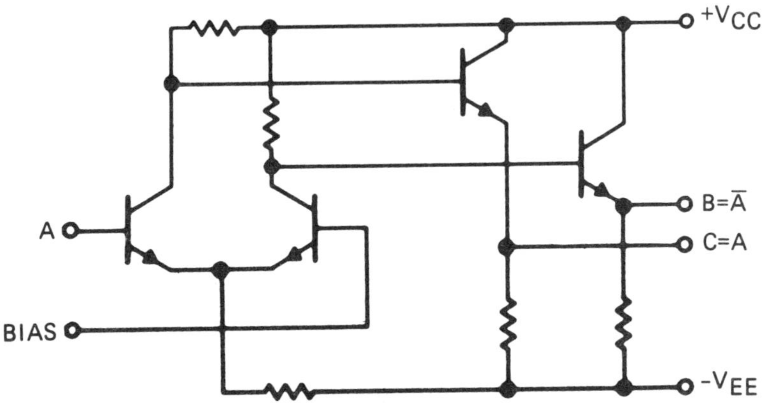

ECL BASIC GATE

FIGURE 3-9

The basic concept of ECL is a differential pair followed by two emitter followers, one for direct output and the other for inverted output. The gating function is provided by paralleling one transistor of the differential pair with one or more additional transistors, collector to collector and emitter to emitter, bringing out each base separately as a gate input. The combination is an OR/NOR gate for positive logic or AND/NAND for negative logic. The bias on the base of the second transistor of the differential pair determines the point at which switching takes place on the other inputs (bases).

ECL logic operates in a nonsaturating mode and hence avoids transistor storage time and its attendant speed limitations as well as the trade tradeoffs required by TTL to attain high speed operation. Output pulldown resistors may or may not be internal to the IC package.

Omitting these resistors permits more flexible arrangements of wire-ORing where two or more outputs are connected in parallel.

A three-input ECL gate is shown in Figure 3-10. Gates are added merely by paralleling the collectors and emitters of additional transistors and bringing out the bases for the additional inputs. The three-input gate of Figure 3-10 has inputs A, B and C, NOR output D, and OR output E for positive logic. This becomes an AND/NAND gate for negative logic. All inputs are negative with respect to ground, but for positive logic the more negative is 0, while for negative logic the more negative is 1. Terminals F and G may be provided for gate expansion.

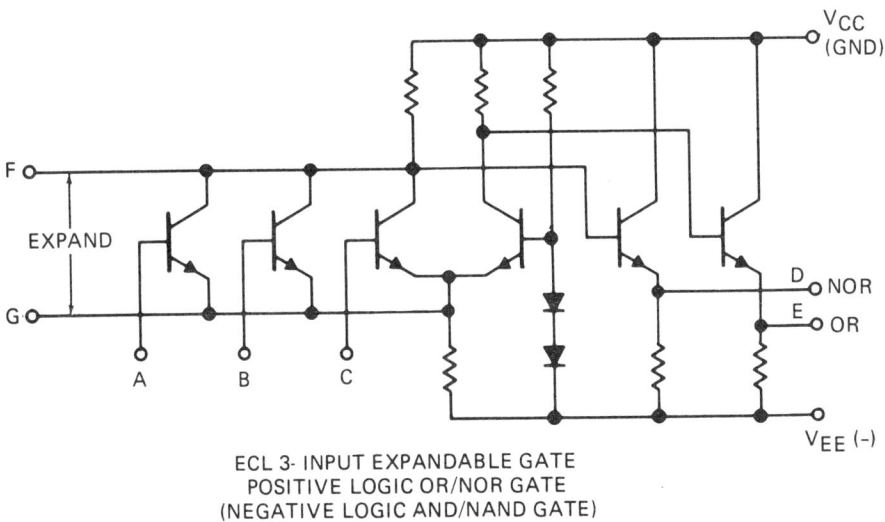

ECL 3- INPUT EXPANDABLE GATE
POSITIVE LOGIC OR/NOR GATE
(NEGATIVE LOGIC AND/NAND GATE)

FIGURE 3-10

The gate expander of Figure 3-11 will add five more inputs to the gates of Figure 3-10 by connecting F' to F and G' to G, making an eight-input gate. Generally unused gates of ECL logic circuits may be left open, which is not a universal rule for IC logic gates. If there is external circuit leakage to any unused gate, it should be connected directly to VEE. Leakage would include high frequency or other AC field pick-up as well as DC ohmic leakage, keeping in mind that the base input circuit has a relatively high impedance, even with pull-down resistors.

Logic which toggles (flip-flop changes state) at rates of over 50

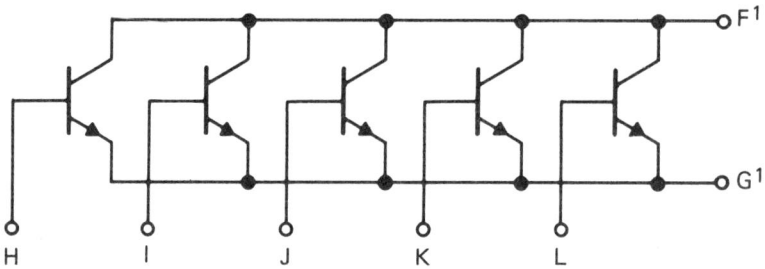

5- INPUT GATE EXPANDER
FOR ECL LOGIC GATES

FIGURE 3-11

MHz and/or has a gate propagation delay of under 6 ns may be considered high speed logic. Propagation delay is defined as the time it takes for the output of a circuit to respond to an input signal. Saturated logic gates such as TTL having storage at both collector-base and base-emitter diodes have difficulty in meeting such a test. However, ECL is operated out of saturation and is easily classed as high speed, attaining toggle frequencies approaching a gigahertz and gate delays of less than a nanosecond.

Motorola has been a leading exponent of ECL, calling their highly developed line by the proprietary term MECL. The line has been expanded into a family of five distinctive groups, variously called MECL I, MECL II, and so on. Figure 3-12 is an interesting tabulation of comparative features of this line. Circuitwise the entire Motorola line is based on the ECL as described above.

High speed logic design, in order to take advantage of the inherent circuit response, requires added precautions which may be relatively unimportant in low or medium frequency systems. Interconnecting wiring may play an important part and may well be the determining factor in the final overall speed of the system. Time delays, distortion due to signal reflections and crosstalk between signal leads must be dealt with. Transmission line type wiring is often indicated. High frequency noise may and must be minimized.

HTL. While 3.6 and 5.0 volts are widely used VCC voltages for ICs, there are applications where higher voltages are advantageous. Logic circuits operating at higher voltages, for example from 7-30 volts VCC, are generally grouped as high voltage logic (HTL). Indus-

MECL FAMILY COMPARISONS

FEATURE	MECL I	MECL II	MECL 10,000		MECL III
			10,100 SERIES 10,500 SERIES	10,200 SERIES 10,600 SERIES	
1. GATE PROPAGATION DELAY	8 ns	4 ns	2 ns	1.5 ns	1 ns
2. GATE EDGE SPEED	8.5 ns	4 ns	3.5 ns	2.5 ns	1 ns
3. FLIP-FLOP TOGGLE SPEED (MIN)	30 MHz	70 MHz	125 MHz	200 MHz	500 MHz
4. GATE POWER	31 mW	22 mW	25 mW	25 mW	60 mW
5. SPEED-POWER PRODUCT	250 pJ	88 pJ	50 pJ	37 pJ	60 pJ
6. TRANSMISSION LINE CAPABILITY	NO	ON SOME DEVICES	YES	YES	YES
7. WIRE-WRAP CAPABILITY	YES	YES	YES	YES	NO
8. OUTPUT PULLDOWN RESISTORS	YES	OPTIONAL	NO	NO	NO
9. INPUT PULLDOWN RESISTORS	NO	NO	50 KΩ	50 KΩ	2 KΩ & 50 KΩ

FIGURE 3-12

trial uses often require high noise immunity while tolerating a relatively slow switching time. Slow response is favorable to noise immunity as well as the relatively high level signals required to switch. In
this form of logic the input voltage required to cause switching at the
output is purposely made high and may be the order of one half of
VCC. Thus, an HTL gate operating at VCC + 14 volts changes state
(switches) at an input voltage of +7 volts. On the other hand, the
output low (0) may be kept as low as in other types, say around 0.1
volt, so that the tolerance or zero range is also very wide (0.1 to 7.0
volts).

   HTL can be provided in a number of circuit types. One form
employs diodes at the input and an emitter follower configuration to
divide the output voltage as shown in Figure 3-13. The zener diode
between the input and output transistors insures a level shift in the
input signal requirement.

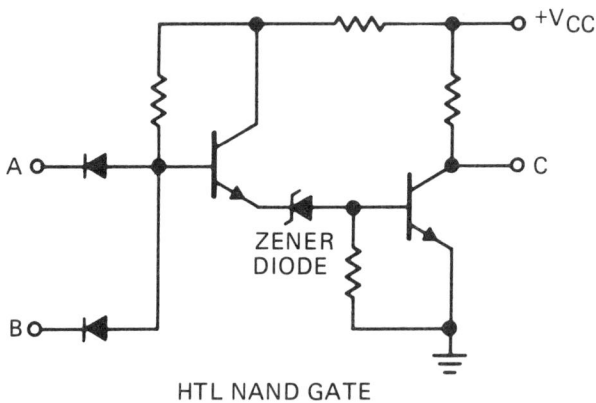

HTL NAND GATE

FIGURE 3-13

   DC UTL. Direct coupled unipolar logic, i.e., logic integrated of
junction field effect transistors, was the first to use field effect transistors. A binary full adder was integrated, resulting in a very simple
pattern of JFETs connected in series by semiconductor bridges. However, more sophisticated forms of field effect devices have crowded
out this early effort.

   MOS/FET. Logic formed of metal-oxide-semiconductor integrated circuits (MOS) is merely the term applied to field-effect transis-

tor logic in which the gate is insulated with a metal oxide such as silicon oxide. The advantage of this construction is high packing density and low power requirements. However, the MOS is slow compared to some other logic forms, and reliability has been a problem. The gate insulator is prone to puncture even from a static charge and requires special handling techniques. Bias and operating voltages may be higher, making this form of logic incompatible with lower voltage forms.

The basic gate circuit is similar to RTL but since the gate can draw only a microscopic DC current, current limiting circuits are unnecessary. Figure 3-14 shows a basic MOS gate. The gates are protected by diodes but these add to the component count and therefore to the package cost. The MOS/FET is a depletion mode device, so that for N-channel V_{CC} is positive and the gate out-off voltage (logic 0) is negative. This dual polarity requires both positive and negative voltage power supplies in the system.

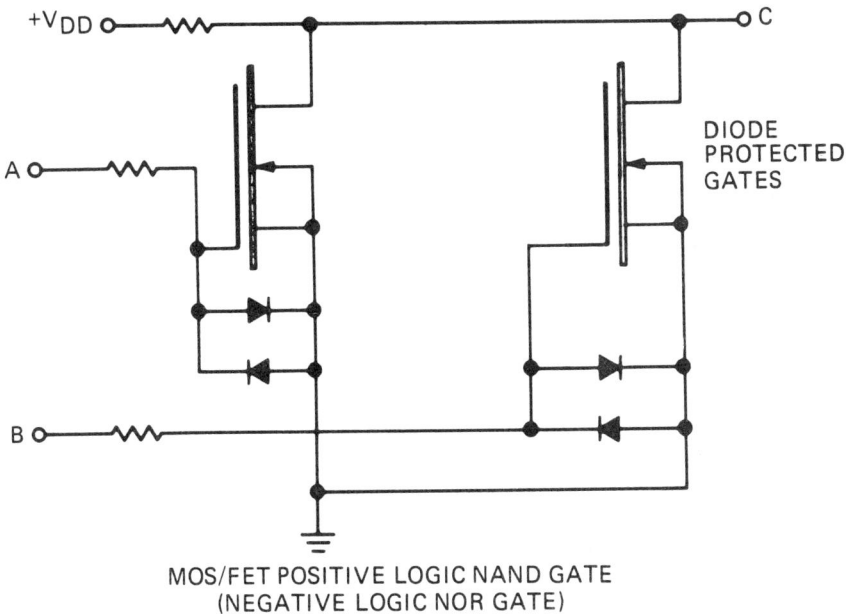

MOS/FET POSITIVE LOGIC NAND GATE
(NEGATIVE LOGIC NOR GATE)

FIGURE 3-14

The higher operating voltages encountered in MOS/FET logic may be reduced by using silicon nitride under the gate, which has a high dielectric constant, improving the gain factor by about 50%.

C/MOS. COS/MOS or C/MOS complementary metal oxide field effect transistor logic (COS/MOS or C/MOS) has some characteristics making it particularly interesting. The C/MOS transistor is an enhancement mode device. That means that an N-channel device with a positive drain voltage requires a positive gate voltage to turn on while a P-channel device operates on negative drain and negative gate voltages. When an N-channel and a P-channel device are connected in series (see Figure 3-15), a gate voltage near V_{SS} turns on the P-channel and turns off the N-channel, providing a high output. Conversely, when the input is made high, close to V_{DD}, the reverse conduction takes place; N-channel conducts and P-channel becomes non-conducting, producing a low (0) output. Since the switching from one condition to the other takes place at a voltage midway between V_{SS} and V_{DD}, the circuit has a high noise immunity in terms of actual noise voltage. However, the very high impedance of the gate circuits makes them sensitive to stray signal pick-up. C/MOS logic is not particularly high speed. When it is necessary to lower the circuit impedance by shunt capacitors, the speed is still further reduced. Lowering the operating voltages also slows response speeds.

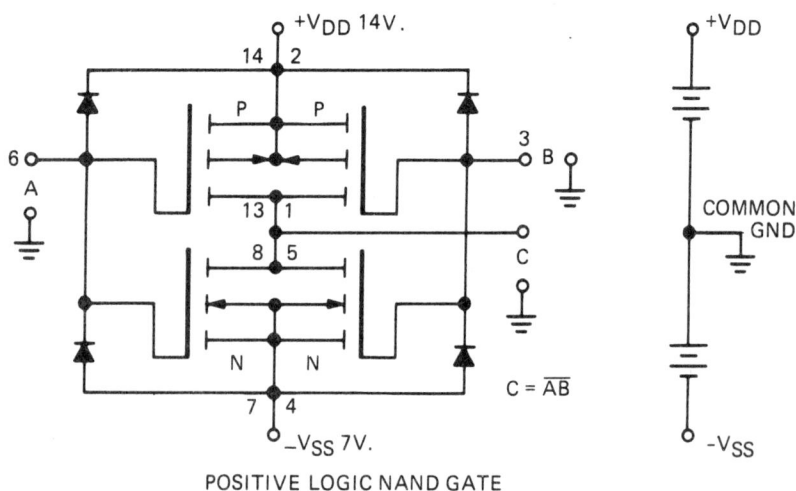

POSITIVE LOGIC NAND GATE

FIGURE 3-15

The C/MOS positive logic NAND circuit of Figure 3-15 and the NOR gate of Figure 3-16 show how basically simple C/MOS gate circuits are. With no load resistors required, and since one transistor in

each gate is always off, the gates operate on very low power. At low
VDD-VSS voltages the C/MOS becomes compatible with other forms
of low voltage logic. At higher voltages the gate delays are reduced,
noise immunity is increased and the device becomes more of an HTL
type circuit.

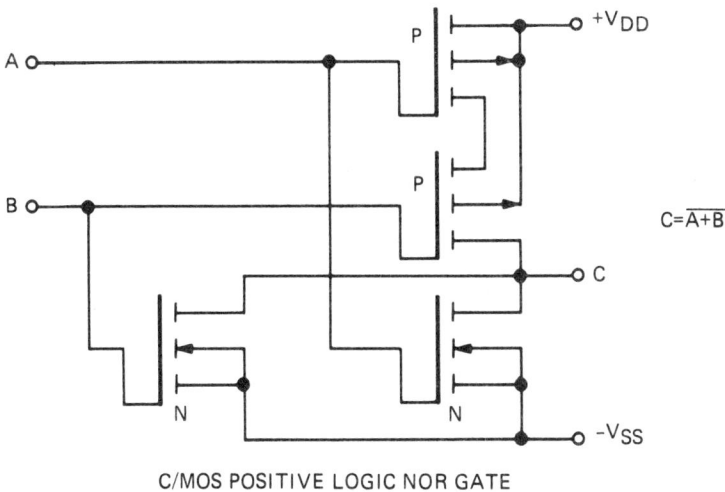

C/MOS POSITIVE LOGIC NOR GATE

FIGURE 3-16

The ground symbol is often omitted from logic diagrams. Ground
will generally be understood to be the common potential of the circuit
or system. For positive logic, an input or output signal 0 means zero
potential with respect to common. More specifically, the zero is anything below, say, +0.4 volt, including going below zero. Furthermore, the logic gate or other device may not have a common or ground
connection as in Figure 3-15. The positive and negative voltages are
specified with respect to ground or common and are such that with
specified 0 and 1 input voltages applied, the output will switch roughly
midway between 0 and 1, thereby insuring switching and providing
maximum noise immunity.

A relatively new and useful concept, that of three-state logic, is
made possible by the C/MOS transmission gate or bilateral switch.
When added to a logic gate, the output can be low (0), high (1) or high
impedance (open circuit). Also the low power requirements of C/MOS

permit high packing densities and many more active circuits per package without incurring thermal problems.

Figure 3-17 illustrates the basic C/MOS transmission gate together with its logic symbol. The gate is turned ON by placing a 0 on gate 1 and a 1 on gate 2. When thus turned ON the gate passes a voltage in either direction provided it lies between VDD and VSS. The gate is turned off by placing a 1 on G_1 and 0 on G_2. This gate is intended for use at the output of a logic device as shown in Figures 3-18 and 3-19. A single 1 on the disable terminal is inverted and supplies 1 to G_1 and 0 to G_2, turning the gate OFF. In figure 3-18 the gate is integrated into the gate, providing a three-state logic gate. With the disable input at 0, the gate passes from input to output as an inverter. With a 1 on the disable input, the normal signal input cannot cause a change in output, i.e., it is in "Don't care" state designated X.

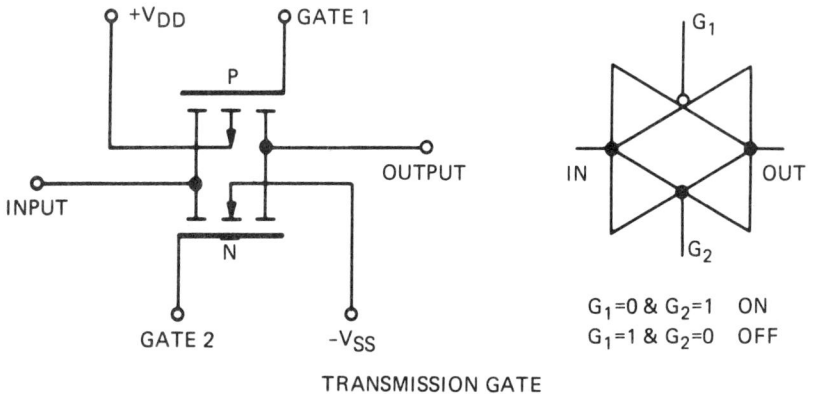

$G_1=0$ & $G_2=1$ ON
$G_1=1$ & $G_2=0$ OFF

TRANSMISSION GATE

FIGURE 3-17

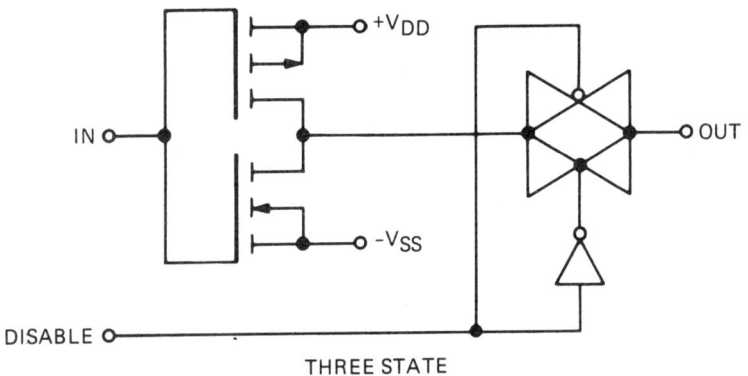

THREE STATE

FIGURE 3-18

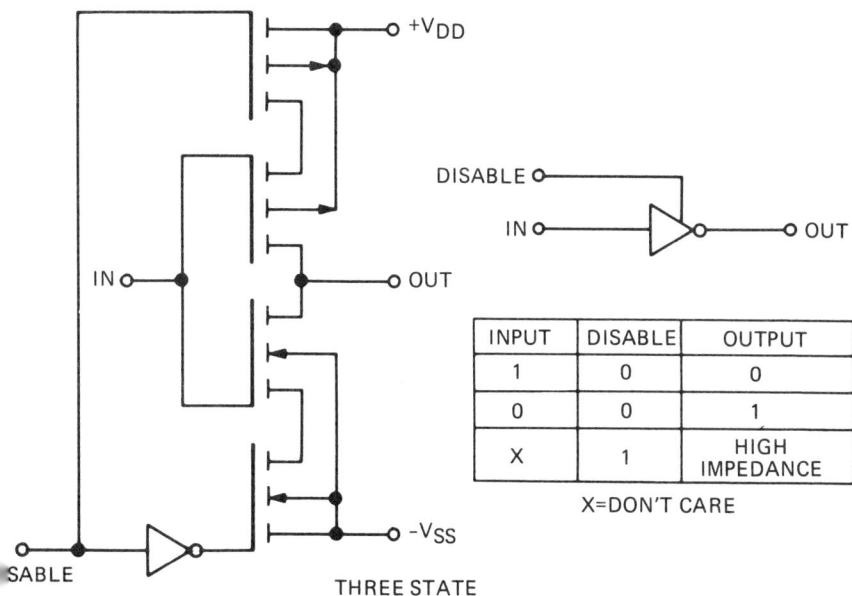

INPUT	DISABLE	OUTPUT
1	0	0
0	0	1
X	1	HIGH IMPEDANCE

X=DON'T CARE

THREE STATE

FIGURE 3-19

An extension of C/MOS is known as SOS (silicon on sapphire). The sapphire provides an electrically insulating substrate so that transistors are grown on insulated islands. This virtually eliminates parasitic capacitances and among other advantages provides higher operating speeds.

Wired-AND/OR gates are available because of transistors.

The way in which the circuit of Figure 3-6 operates has been explained previously in connection with Figure 2-13 in Chapter 2. However, it brings up an important point regarding interconnection of basic digital IC gates. Most complex IC combinations are made of cascaded gate circuits. The basic gates have been designed as primarily intended to be cascaded. What happens when parallel connections are made? It is necessary to look at the actual or equivalent circuitry of the gate to decide. Inputs can generally be connected in parallel, particularly when they are positively driven to 1 or 0. As to outputs there can be no generalization. The DTL logic of Figure 3-5 when the outputs of two two-input NAND gates are connected in parallel becomes a four-

Thus, the typical TTL gate cannot be connected with paralleled outputs. Before attempting the paralleled output (SIRED), it should be determined whether or not it is a permitted connection for the particular gates involved.

Complex logic circuits are built from simple basic gates.

Only the basic logic gates have been described so far. Actually these basic circuits are repeated and combined in various ways to form the flip-flops and most other complex forms. Understanding the basic forms prepares the way to understanding the complex forms. Chapter 4 will describe the construction and operation of various types of flip-flops. Just as the NAND and NOR gates are the building blocks of the flip-flops, the flip-flops are the building blocks of the shift registers, latches, counters and dividers to be described in Chapter 5.

4

How Flip-Flops
Are Formed and Used

Defining the flip-flop.

A circuit which can exist indefinitely in either one of two possible states provides a storage cell for a bit of information. The storage condition is yes or no, 1 or 0, etc. Once the information bit is stored, the cell maintains the set condition until operated on by external means. If two transistors are directly cross-connected, input of one to the output of the other, they form a binary storage device more commonly known as a flip-flop. Figure 4-1 shows a simple flip-flop made up of two transistors cross-coupled so that the output of one is always the complement of the output of the other. The inputs to the transistor

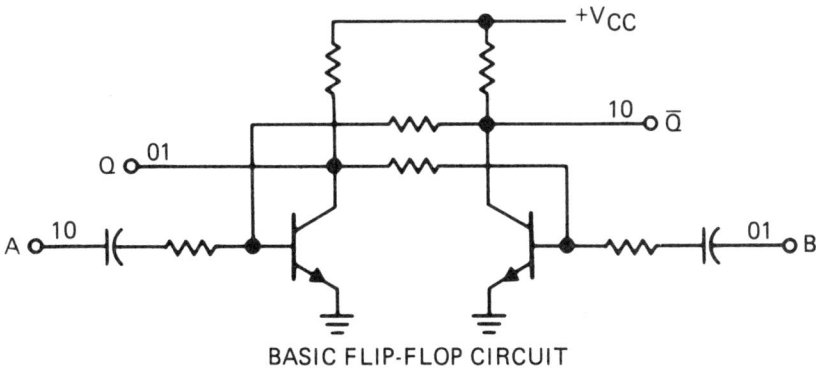

BASIC FLIP-FLOP CIRCUIT

FIGURE 4-1

bases are labeled A and B while the outputs are labeled Q and $\overline{Q}$. Thus, if input A is made high (1), output Q at its collector will be inverted or low (0), and with B high (1) output $\overline{Q}$ will be low (0). With two inverting stages cross-coupled as shown, the output of one stage will be in the same state as the input of the other stage. The states of input B are the same as the states of output Q and the states of input A are the same as the states of output $\overline{Q}$. Since these latter like pairs are coupled together, the circuit will be stable and will hold a given state indefinitely (as long as VCC bias is applied). To change the states it is merely necessary to change the state of one of the inputs momentarily.

Starting wtih the simple basic circuit described above, flip-flops have been integrated in many forms. Two inverting amplifiers cross-connected as shown in Figure 4-2 are the integrated equivalent of the two-transistor circuit of Figure 4-1. However, a more useful form is that shown in Figure 4-3 where two NAND gates are cross-connected and separate inputs (R and S) are provided. The resulting truth table is shown in Figure 4-4. Note that when both R and S inputs are made high (1), the resulting output depends on which went high first: if S, the Q output stays high and Q goes low; if R, the inverse condition results.

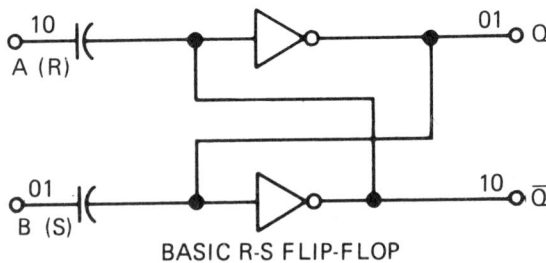

BASIC R-S FLIP-FLOP

FIGURE 4-2

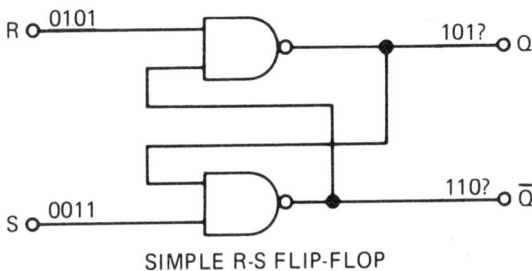

SIMPLE R-S FLIP-FLOP

FIGURE 4-3

S	R	Q	$\overline{Q}$
0	0	1	1
0	1	0	1
1	0	1	0
1	1	FIRST "1" CONTROLS	

R-S FLIP-FLOP TRUTH TABLE

FIGURE 4-4

Changing the input designations from A and B to S and R was deliberate. The S and R designations are widely used to denote "Set" and "Reset" respectively. Thus, we have a basic form of R-S flip-flop employing two NAND gates, even more basic IC devices. The most important thing about a flip-flop is that it "remembers." The cross-connections of output to input force it to hold its set or reset state after the set and reset signals are removed. Figure 4-4 is the truth table for the R-S flip-flop.

There are a number of uses for the simple R-S flip-flop, and it should be remembered that R-S flip-flops are basically simple, requiring few components, and therefore less expensive than other types. One use is in cleaning up a "dirty" signal. Signals containing noise components are very troublesome in many circuits. A dirty or noisy signal as shown at A in Figure 4-5 triggers the R-S flip-flop, providing a clean regenerated signal B on its Q output. Of course the noise must not exceed the noise threshold of the flip-flop if a true output is to be obtained. An example of a dirty signal is one generated by a manually operated push-button where contact bounce produces an indefinite closure point.

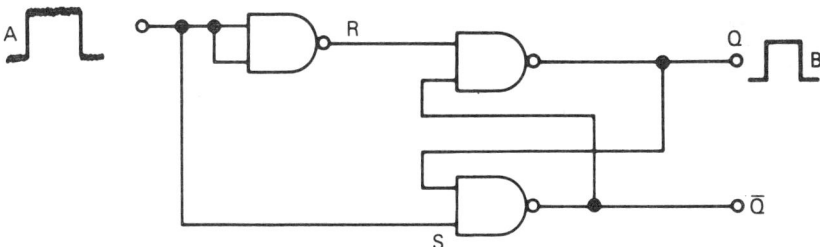

SQUARE WAVE REPEATER

FIGURE 4-5

Industrial control systems use the simple R-S flip-flop or its equivalent, as shown in Figure 4-6, for switching purposes. With two push-buttons the Q output may be enabled or disabled. When the S push-button is pressed, grounding the S input, the Q output goes high as shown in the truth table. Similarly, pressing the R push-button sends Q low and Q̄ high. The action is quite analogous to that of a latching relay, and at one time this circuit was popularly called a *latch*. Latching is instantaneous, typically a few nanoseconds, and the circuit will hold as long as bias power is applied to the flip-flop. However, if power is interrupted and then restored, the flip-flop will come back on in a random fashion. The circuit must be set all over again.

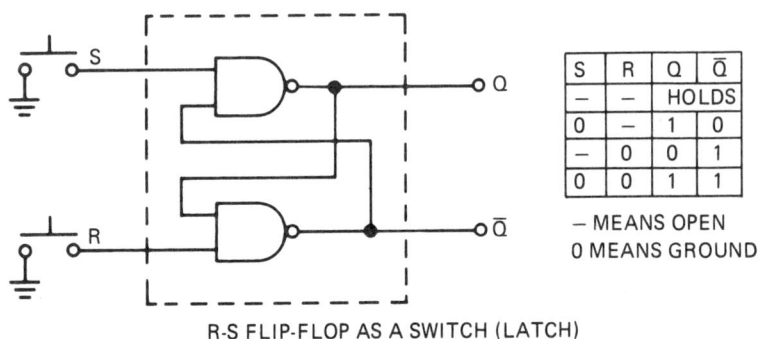

S	R	Q	Q̄
–	–	HOLDS	
0	–	1	0
–	0	0	1
0	0	1	1

– MEANS OPEN
0 MEANS GROUND

R-S FLIP-FLOP AS A SWITCH (LATCH)

FIGURE 4-6

Synchronous flip-flop operation is required in some cases.

Circuits and systems operating at high rates of speed practically demand synchronous, that is, clocked, operation. This is due to the fact that generally many things are happening at the same time and if they are not all synchronized, they cannot be made to coordinate or cooperate. The clock or synchronizing means is a square wave generator often crystal controlled which provides a source of uniformly spaced pulses. These pulses are distributed throughout the system and to whichever circuits require synchronizing. These signals are called "clock" pulses. They must have a shape compatible with the logic used in the system. That is, they must have a rise time, duration, and fall time within defined limits or malfunction of the circuits clocked may result.

There are a number of reasons for clocking a flip-flop. If data is being passed through the flip-flop, it must be enabled at the time when the desired data to be passed is at its input. The clock pulse enables the flip-flop. A simple and obvious method of using the clock pulses is to apply them to one input to a NAND gate with the data applied to the other input. Such a circuit is shown in Figure 4-7. Only when the clock pulse is high can data on input D be passed to the R-S flip-flop for storage or other use. The operation of clocked circuits is often illustrated in graphical form as shown in Figure 5-8. The D line shows a series of data pulses reflecting a changing data state on the line; the C line shows the regular clock pulses; and the Q line shows the resulting output. Assuming the change of state takes place on the leading edge of the clock pulse, this is called an "edge triggered" circuit. Both leading edge and trailing edge triggering is used in logic circuits. The trailing edge, for example, is used in master-slave flip-flops where the leading edge sets up the master and the trailing edge locks out further input changes and transfers the data to the slave flip-flop.

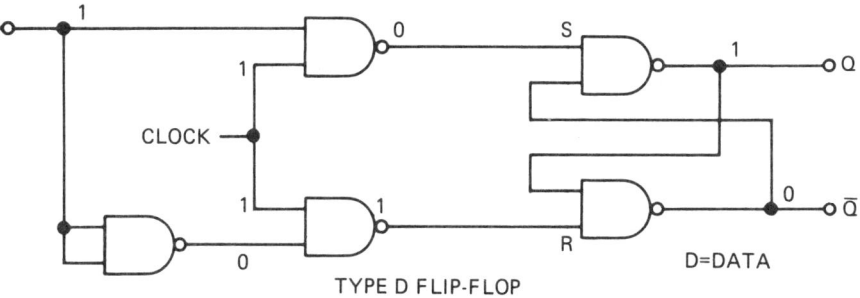

TYPE D FLIP-FLOP

FIGURE 4-7

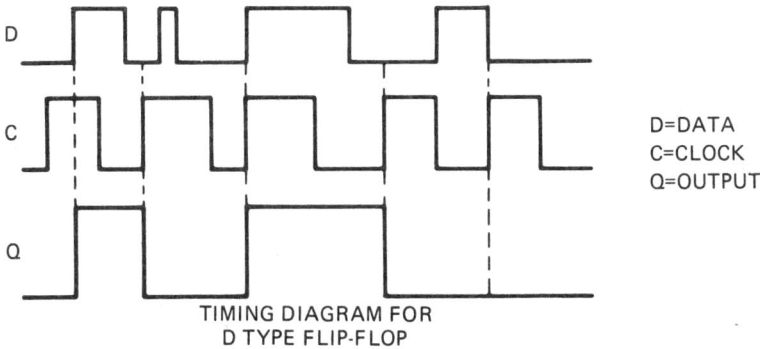

D=DATA
C=CLOCK
Q=OUTPUT

TIMING DIAGRAM FOR
D TYPE FLIP-FLOP

FIGURE 4-8

The toggle flip-flop (R-S-T) is one simple form.

A toggle flip-flop (R-S-T) is one which changes state upon receipt of each clock pulse. This function is useful in counters, timers, frequency dividers and the like. A self-steered RS flip-flop can be made to toggle, that is, change its output state once for each clock pulse. Such an arrangement is shown in Figure 4-9. The toggle was a natural in vacuum tube type flip-flops with symmetrical capacitor inputs supplied with a timing pulse. The pulse turned the ON side OFF and the OFF side ON. However, the flip-flop constructed of integrated gates as shown in Figure 4-9 has no such natural disposition to switch states. Unless special precautions are taken to see that the circuit is well balanced and that the clock pulse meets certain waveform requirements, the RST toggle is an erratic and unpredictable device.

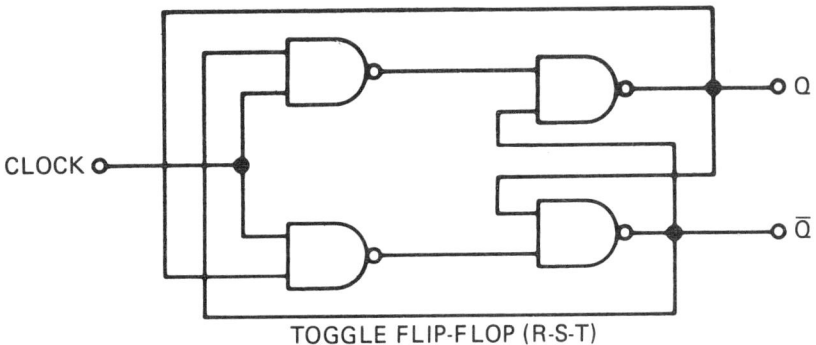

TOGGLE FLIP-FLOP (R-S-T)

FIGURE 4-9

The J-K flip-flop has the most versatile capabilities.

Substituting three-input NAND gates for the two-input ones shown in Figure 4-9 results in what is known as a J-K flip-flop, shown in Figure 4-10. Broadly defined, a J-K flip-flop is a clocked flip-flop having J and K inputs which may be operated to control the Q and Q̄ outputs but which has the further property that whatever inputs are supplied to those J and K inputs, the output conditions are predictable (as contrasted with the simple R-S flip-flop having ambiguous conditions of output). Another property of the J-K flip-flop is that if J and K

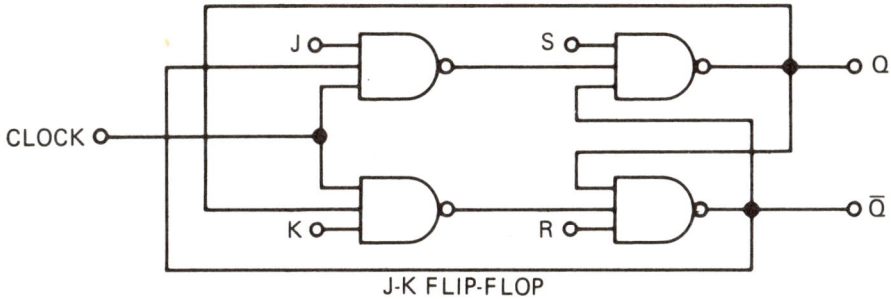

J-K FLIP-FLOP

FIGURE 4-10

inputs are both made high, the outputs will toggle with a pulsing clock input.

The J-K flip-flop is very versatile and can be provided by a number of gate combinations. The various forms have differing characteristics and are provided to meet differing circuit or system requirements. It is probably the most universal type of the entire flip-flop family. The origin of the term J-K is obscure but there is apparently rather good agreement as to what it denotes. Accordingly, a J-K flip-flop is generally defined as ''a flip-flop having in addition to R and S inputs, which take precedence under certain operating conditions, also a clock input and two additional inputs known as J and K, which operate in conjunction with the clock, and which can produce only predictable outputs.''

Other flip-flops have been devised.

It will be seen from the above that flip-flops can be formed from various combinations of NAND gates. They can also employ NOR gates and inverters. The various manufacturers of ICs have many different concepts of how flip-flops should be made. Each concept has a reason behind it. The uses to which flip-flops are put do require some variety. Also it would not be economical to employ a complex form of flip-flop when a simple form would perform the required function. While some forms are generally recognized by all manufacturers and are designated by generally accepted terms, other forms are made with a given manufacturer's own unique designation.

Some practical ways to use flip-flops.

Just as gates may be used simply as gates or may be combined to form more complex devices such as flip-flops, so flip-flops may be combined to form even more complex devices. As a matter of fact, the flip-flop is a key building block of a wide range of important IC devices. These combination devices include counters, dividers, signal regenerators, memories, registers, converters, delays, clock waveform generators and many even more complex devices. Devices formed by combining flip-flops will be discussed in Chapter 5.

Pulses and synchronous operation as applied to flip-flops.

The logic gates and simple flip-flops have been described as being operated by changing input DC levels. Many simple control systems are operated on this basis. However, as systems become more complex, the DC mode inhibits the flexibility of design and cannot be tolerated. The broad spectrum of IC applications is possible due in large part to the use of pulses, clocks and synchronous operation. Along with the expanding art of the IC is the development of synchronous system techniques.

The use of pulses in place of DC levels provides great freedom in system design as well as additional IC functions. Frequency division can only be accomplished by means of pulse response circuits. Frequency synthesizers can only be designed by using pulse responsive circuits. Counters, timers and practically all complex systems such as digital computers are pulse operated. Thus, in addition to the DC level response to IC gates and gate combinations it is necessary to know the pulse response.

In addition to pulse response, most pulse operated systems must also be synchronous. For example, a NAND gate responds to DC level changes and it also responds to pulses of the proper shape. However, if the pulses to be NANDed do not arrive at the NAND gate inputs at the same time, i.e., in synchronism, the output is not changed as intended. If the information to be NANDed is not generated simultaneously, i.e., synchronously, some means must be employed to bring it into concert. This may be done, for example, by some temporary storage device such as a simple flip-flop. Then, at a predetermined instant a clock pulse enables a gate, and the stored information is transferred to a synchronized portion of the system. Since there is no such thing as instantaneous operation of a device or a system, certain timing requirements are significant.

One of the important characteristics of an IC in a synchronous system is the propagation delay. Another is the transition time. Figure 4-11 shows an input waveform and a resulting output waveform, idealized for purposes of explanation. The upper wave goes from 0 to 1 and back to 0. The resulting output, assumed to be inverted as would be the case for a NAND or NOR gate, goes from 1 to 0 and back to 1.

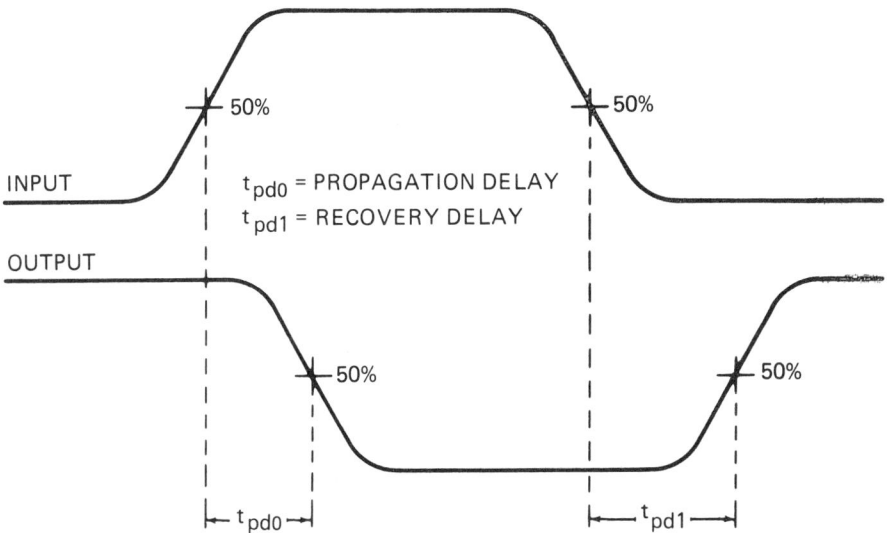

FIGURE 4-11

It is conventional to designate the time lapse from the time the input signal passes the 50% point to the time the output passes the 50% point with the input going from 0 to 1 as the propagation delay for an output changing from 1 to 0 as tpdo, and the recovery delay as tpdl. The sum of these two is called pair delay and is thus equal to tpdo + tpdl.

Another characteristic of such circuits is the transmission time delay defined as follows in conjunction with Figure 4-12. The first, tdo, is defined as the time required for the output to change 10% of its transition from 1 to 0 with respect to the 10% change point in the input;

second, tdl is the time required for the output to change 10% of its transition from 0 back to 1 with respect to the causative 10% change point in the input; third, t_0 is the time required for a 90% to 10% change in output in going from 1 to 0; and fourth, t_1 is the time required for a 90% to 10% change in output in going from 0 to 1.

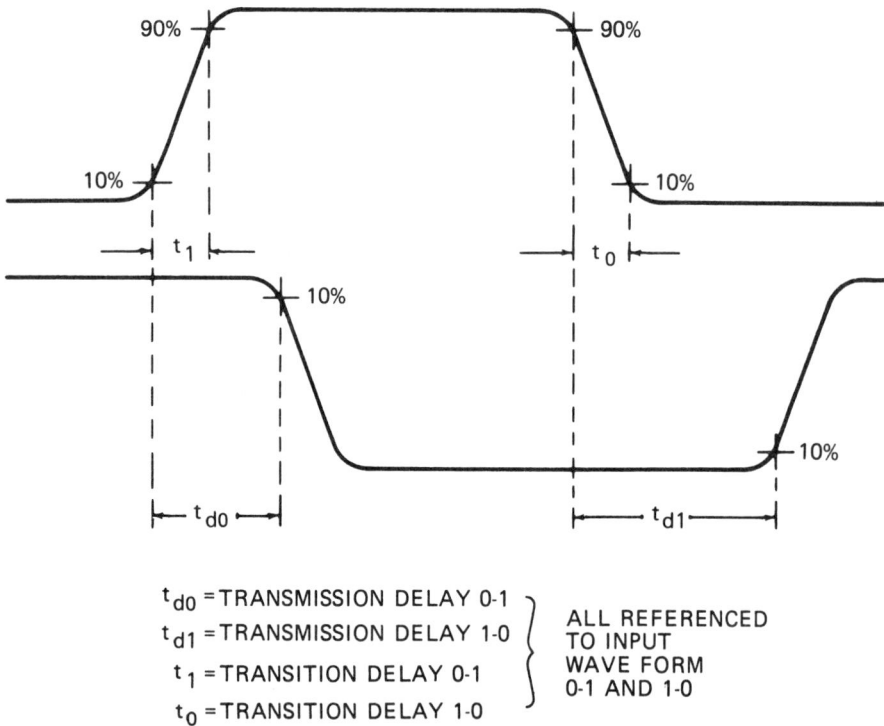

t_{d0} = TRANSMISSION DELAY 0-1
t_{d1} = TRANSMISSION DELAY 1-0
t_1 = TRANSITION DELAY 0-1
t_0 = TRANSITION DELAY 1-0

ALL REFERENCED
TO INPUT
WAVE FORM
0-1 AND 1-0

FIGURE 4-12

There are still further time elements of importance in pulsed synchronous systems. One is set-up time, i.e., the time during which a signal must be present at a given gate before it can be utilized. Another is clock pulse width, i.e., the time the clock pulse must be high, or low as the case may be, to properly utilize the presented information. Still another is settling time, i.e., the time which must be allowed for circuits to stablize before the application of another clock pulse.

Clocking the system a "must" in many applications.

The terms "clock" and "clock pulse" have been used here without sufficient introduction to the meaning of these terms. It has been stated above that pulsed systems generally have to be synchronous systems. The means of achieving synchronization is the clock or the clock pulses. The clock is a pulse generator of accurately timed pulses, generally crystal controlled. All pulse operated portions of a given system are clocked, i.e., supplied with a synchronizing control by means of these clock pulses. In other words, all operations are controlled and are in unison by virtue of being gated or enabled by the clock pulses. Obviously, the higher the clock frequency the more the information which can be handled. On the other hand the clock frequency is limited by the speed of response of the ICs used in the system and how they are combined. Problems multiply at high clock frequencies, as will be discussed later on.

To a first approximation the clock period must be greater than the sum of the propagation delays and the set-up time. The propagation delay involved in this determination is the longest propagation delay encountered in the system between any two clock points. For example, suppose the longest delay in the system occurs where two flip-flops are coupled through three series connected gates as shown in Figure 4-13.

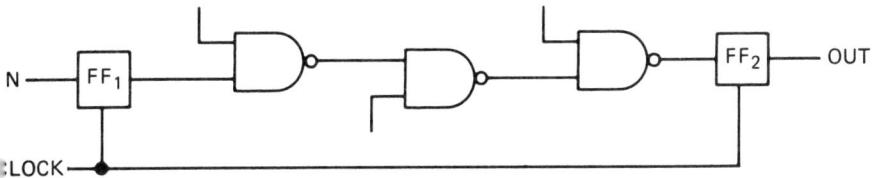

FIGURE 4-13

The propagation delay here is the propagation delay of a flip-flop plus three times the propagation delay of one NAND gate. The set-up time is the time during which the input to the flip-flop must be static before the clock pulse is applied. Suppose, for example, the propagation delays are flip-flop 6 nanoseconds, NAND gates 2 nanoseconds each (6 ns total), and set-up time 8 nanoseconds, or a total of 20 nanoseconds. The clock frequency for such a system must be less than

50 megacycles. (f = 1/[6+6+8]10^{-9} = 50 megacycles.) Since propagation delays and set-up times vary from unit to unit, with supply voltage and with operating temperature, ''worst case'' figures should be used to compute the clock frequency.

Clocked systems vary all the way from a simple local device with a local clock used to keep an orderly process going to very diverse and complex systems. Where an overall systems clock is used along with one or more local clocks which may be of different base frequencies, synchronization of the local clock to the system clock may be necessary. Figure 4-14 shows one possible such system. The local system operates from a 1 MHz clock but receives data at 1/100 that rate, i.e., at 10 KHz. The local clock oscillator is synchronized with the incoming data clock but at 100 times the frequency. This provides a system where data and control at two widely different rates can be handled. One important advantage of a system of this kind is that noise and drop-outs in the incoming data do not upset the system since the high frequency oscillator has a fly-wheel effect bridging over these irregularities.

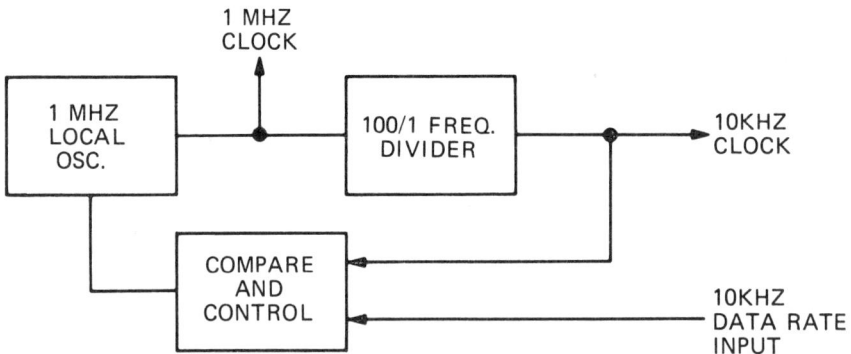

FIGURE 4-14

The simple clock puts out an equally spaced series of pulses. It is often advantageous to make these pulses as narrow as possible and still wide enough to fully enable the circuits they are timing. One of the main reasons for making the clock pulses narrow is to discriminate against noise. If a circuit is enabled only 10% of the time, it may be considered to provide a 10-1 discrimination against noise which is a randomly distributed signal. Another reason is that short clock pulses allow a relatively longer time to set up the circuits to be clocked.

The series of pulses forming clock pulses as described above are called single-phase clock pulses. Some systems benefit from a two-phase clock. For example, suppose a series of signals have been just clocked through a system and an output result is required. The next clock pulse may take place before the system has settled and hence would gate out a transient signal. An inverted clock pulse (a second phase) would provide an additional half clock period to produce the output answer. There are other uses, as well, for an inverted or second phase clock pulse series.

5

Combining Gates and Flip-Flops for a New World of Capabilities

How serial counters are formed of cascaded flip-flops.

A J-K flip-flop with both J and K terminals high changes state each time its clock input goes from 1 to 0. Cascaded J-K flip-flops provide a simple binary counter, as shown in Figure 5-1. Here four flip-flops are connected in cascade by connecting the Q output of one to the clock input (C) of the one following. Read out of the count may be obtained as by lamps connected to the Q outputs. Typically the state of a given flip-flop changes when its clock input goes from 1 to 0 as shown by the arrows. When the first input pulse (1) goes from 1 to 0, the Q_1 output goes to 1, lighting the lamp reading 1 and thereby reading count 1. When the second input pulse goes to 0, Q_1 goes to 0 and Q_2 goes to 1, lighting the number 2 lamp and indicating a count of 2. When the third input pulse goes to 0, Q_1 goes to 1, lighting lamp 1 again. Lamp 2 remains on and the count is read $1 + 2 = 3$ and so on counting in binary fashion as high as one cares to go.

This circuit may also be considered a frequency divider with division factors of 2, 4, or 8. Thus, an output taken from Q_1 has half as many transitions as the input; from Q_2 one quarter as many; from Q_4 one-eighth as many, and so on. Counters and frequency dividers are basically the same thing but with different uses of the resulting outputs.

Counters can divide frequency by almost any factor.

By feeding back at various points and in various ways, counting and dividing circuits can be modified to provide almost any desired

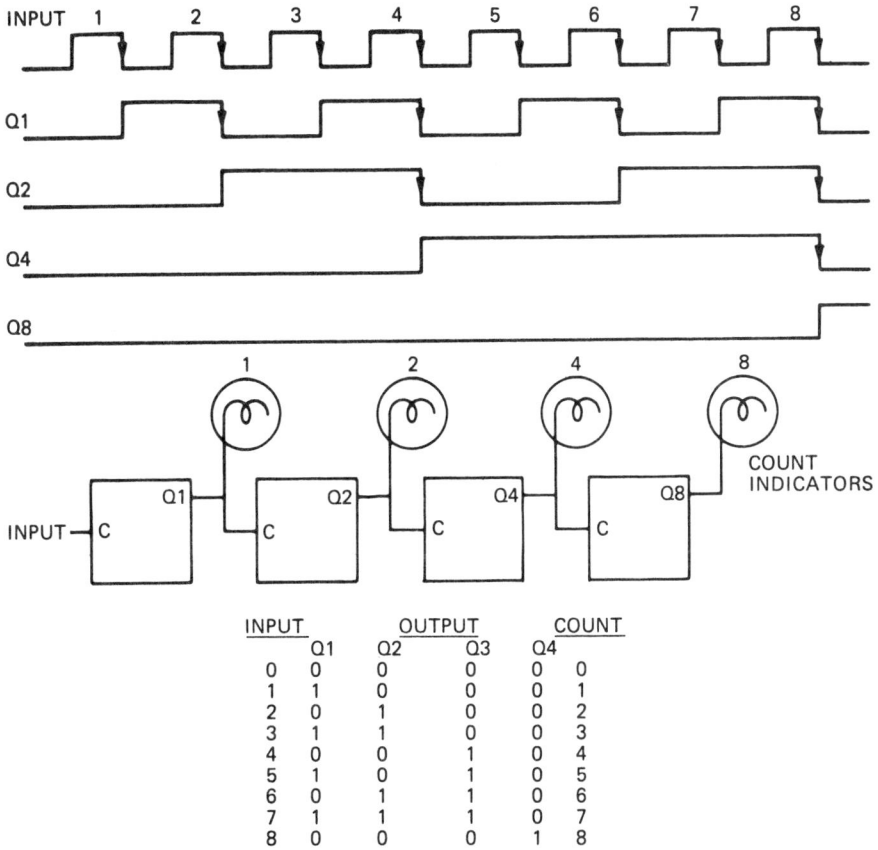

INPUT		OUTPUT		COUNT	
	Q1	Q2	Q3	Q4	
0	0	0	0	0	0
1	1	0	0	0	1
2	0	1	0	0	2
3	1	1	0	0	3
4	0	0	1	0	4
5	1	0	1	0	5
6	0	1	1	0	6
7	1	1	1	0	7
8	0	0	0	1	8

FIGURE 5-1

number or ratio. While there are many ways of modifying a binary counter to cause it to count other than its binary full count, one general method is to force the counter to go from its N-1 state (N being the desired count) to reset to its initial all-zero condition on the Nth count. Another method is to sense the N-1 count, and on the next count to inhibit all stages normally changing to 1 and to force all stages at 1 back to 0.

Three cascaded flip-flops can be used to divide by 2, 3, 4, 5, 7 or 8 as shown in Figure 5-2. Without feedback the first Q output Q_1 divides by 2, the second by 4 and the third by 8. Now on the third count after reset, the Q_1 and Q_2 outputs will both be 1. If they are connected to a NAND gate and the output of the gate is connected to the reset inputs (R) of the flip-flops by closing switch S_1, every third count will reset the circuit and it will divide by 3. On the fifth input pulse Q_1, Q_2 and Q_3 will all be 1 and these NANDed and connected back to reset Rs by closing switch S_2 will cause the divider to divide by 5. Similarly, on the seventh input pulse Q_1, Q_2 and Q_3 will be at 1, and when NANDed and fed back by closing switch S_3 will force the counter to divide by 7. In a like manner a divider/counter may be forced to divide/count by any predetermined number provided one has enough stages of cascaded flip-flops and a NAND gate with enough inputs to combine 1's representing unique Q or $\overline{Q}$ states at the desired count.

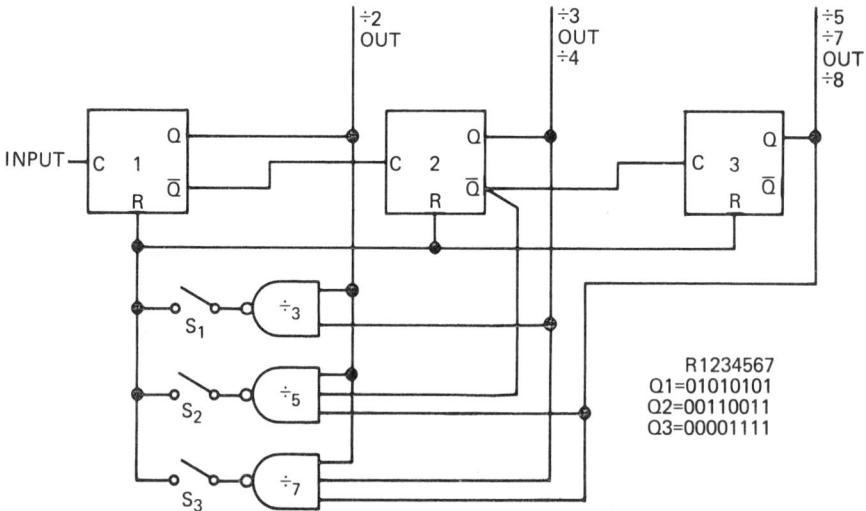

FIGURE 5-2

To count by large numbers the first step is to factor the number. Any multiples of 2 are provided by a single flip-flop, 3 by two flip-flops and feedback as shown above and so on. For example, to count by 60's, factor 60 into $4 \times 3 \times 5$, which can be provided by 2 stages of 2, one stage of 3 and one stage of 5 or a total of seven flip-flops as shown in Figure 5-3. This can be simplified, however, by using 4×15

as shown in Figure 5-4, requiring a total of only six flip-flops.

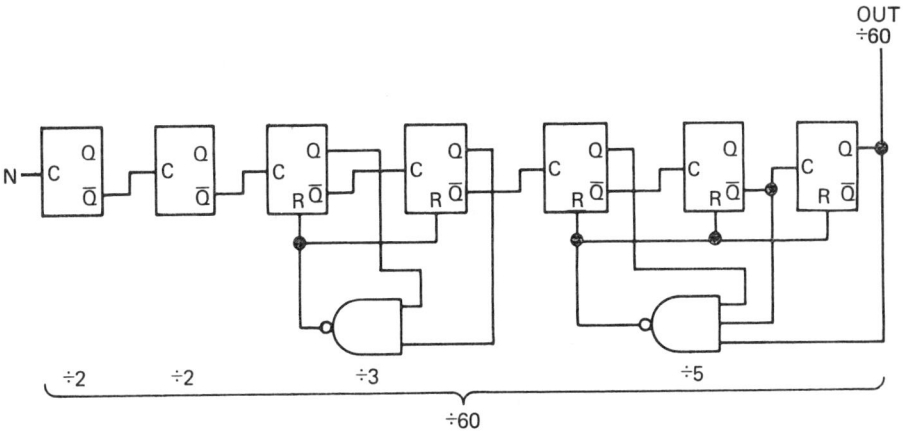

FIGURE 5-3

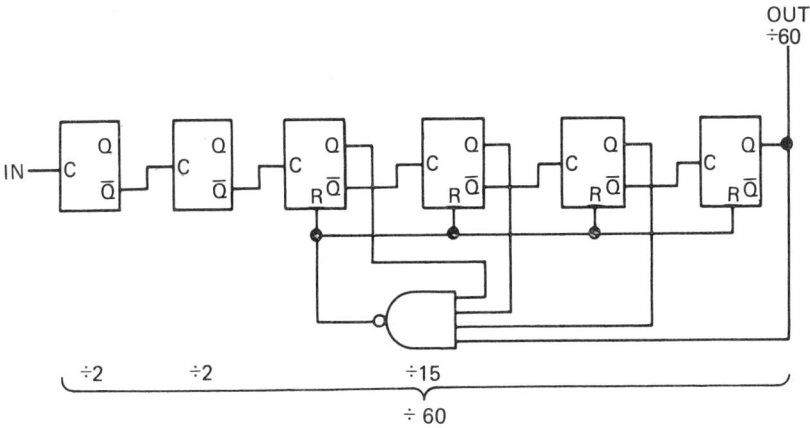

FIGURE 5-4

One widespread use of counters for frequency division is in clocks and watches using high frequency crystals as their frequency standard. These crystals are small, inexpensive, and can be cut to exhibit a high degree of frequency stability. The crystal frequency is divided down to seconds, minutes and hours by a series of cascaded dividers (counters).

One of the conditions which must be met in such a system is that in any portion of the divider chain where feedback is used to modify the normal binary count to some other count, the pulses being counted must be long enough in time to allow the feedback to function during one pulse period.

When the input to a divider is the output of a crystal oscillator or other very high frequency source, it may be necessary to divide the first few times in the chain with very high frequency responsive flip-flops. At some point the frequency will be reduced to a degree where low frequency and generally less expensive flip-flops can be used.

Counters are used to provide interval timers.

There are many applications which require timing a predetermined interval. Here a source of pulses of known frequency or repetition rate is counted until a predetermined count is reached when the counting is stopped. Consider a phototimer as an example. Let the interval be five minutes and the source of pulses be an R-C oscillator putting out ten pulses per second. There are $5 \times 60 = 300$ seconds in the five-minute interval and $10 \times 300 = 3000$ pulses. Hence, the counter is set to count 3000 and signal the end of the counted interval.

Flip-flops can be interconnected to form shift registers.

If JK flip-flops are cascaded as shown in Figure 5-5, the state of any given flip-flop is passed on to the next succeeding flip-flop upon each clock pulse. The input state is passed into the first flip-flop on a given clock pulse and is passed along to each succeeding flip-flop in order upon each succeeding clock pulse. At each nth clock pulse, n being the number of stages in the register, the first in information is started out of the nth flip-flop to the output. Thus is formed a so-called shift register.

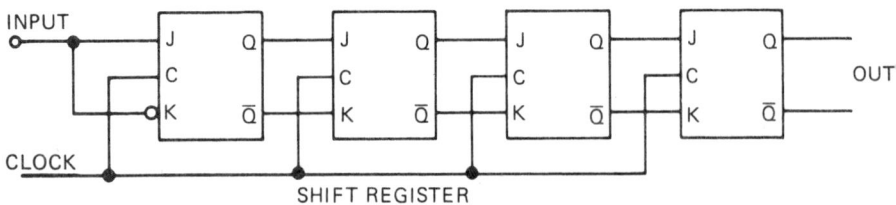

SHIFT REGISTER

FIGURE 5-5

The shift register has many applications; one of the more common is as a temporary storage or memory device. Suppose, for example, four bits of coded information are available at a given time but cannot be used at that same time. These four bits of information can be stepped into a shift register and held there until they can be used. When the time for their use comes, the four bits are stepped out of the shift register.

Another application is where coded information becomes available at one rate but must be used at a different rate. The information can be stored in a shift register under control of a clock having one stepping rate and fed out under the control of a clock having a different stepping rate.

Flip-flops connected another way become ring counters.

The simple serial counter without feedback is known as a ripple counter. Pulses enter the counter, and if it fills, the counting starts all over again. The ring counter is very much like a shift register in that clock pulses are fed to all stages. However, the output of the ring counter (Figure 5-6) is fed back to provide its input. In this way the states of the flip-flops are not only shifted stage by stage but are fed back and recirculated. Thus, with four stages a given pattern is repeated in response to every fourth clock pulse.

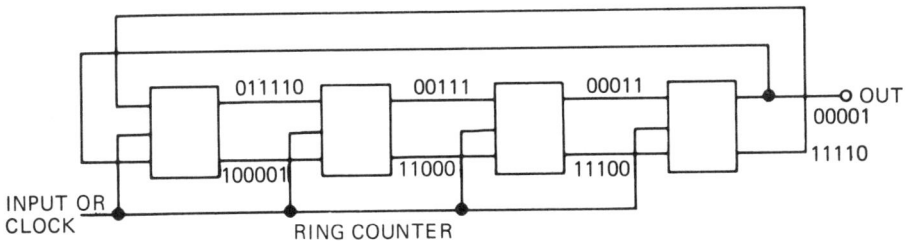

FIGURE 5-6

Large numbers of flip-flops are interconnected to form random access memory (RAM).

One of the most useful and sought after devices in IC technology is the random access memory or RAM. As the name implies the RAM is a memory in which information may be stored and withdrawn in a time interval which is independent of the memory cell address. The access time and the total storage capacity are the two most important characteristics of RAM devices. The RAM composed of IC flip-flops or other IC storage devices has one advantage over all other storage media in that it is compatible with IC logic and requires a minimum of interfacting devices and circuitry. To characterize the RAM using IC technology, they are fast access but of limited storage capacity. Capacity of course can be multiplied indefinitely but must be rated in terms of cost per bit and space requirements.

The actual space per memory cell in the IC memory is greatly expanded in the final form by the packaging which is necessary. LSI techniques provide the best space factor available in IC memories but space is still a final limitation. Still, RAMs made by IC technology are very important products and are assuming ever-increasing functions in logic implemented systems.

Figure 5-7 is a schematic circuit diagram of three bipolar storage cells comprising a TTL type flip-flop. It is simple, fast and reliable. The circuit indicates how the cell is used to write, store and read digital bits. Depending on whether a 0 or 1 is stored, one or the other of the two transistors is conducting. The memory array is produced by columns sharing the digit lines and rows sharing a common word line. With the digit lines at a higher potential than the non-energized word line, corresponding emitters are reverse biased and do not conduct. When the word line potential is raised, either digit line emitter will conduct current to the digit line. Stored data is sensed by detecting this current flow. Writing is accomplished by raising the word line potential and lowering a digit line, forcing the flip-flop to assume the corresponding state. As is the case in practically all IC technology, there are variations of the basic curcuit. Schottky diode, alternate gating means, and designs for minimizing power consumed in unselected flip-flops are among these variations.

Figure 5-8 shows one form of MOS storage cell with transistors used as load resistors. Transistors acting as resistors are more adapted to integration than actual resistors. Two transistors are used as gates activated by the word line and enabling access of the digit lines. The state of the cell is determined by raising the word line to turn on the gates. Writing is done in a similar manner, i.e., by raising the word line and setting up the write logic on the digit lines. When not being accessed, the cell may be switched to a low power standby mode.

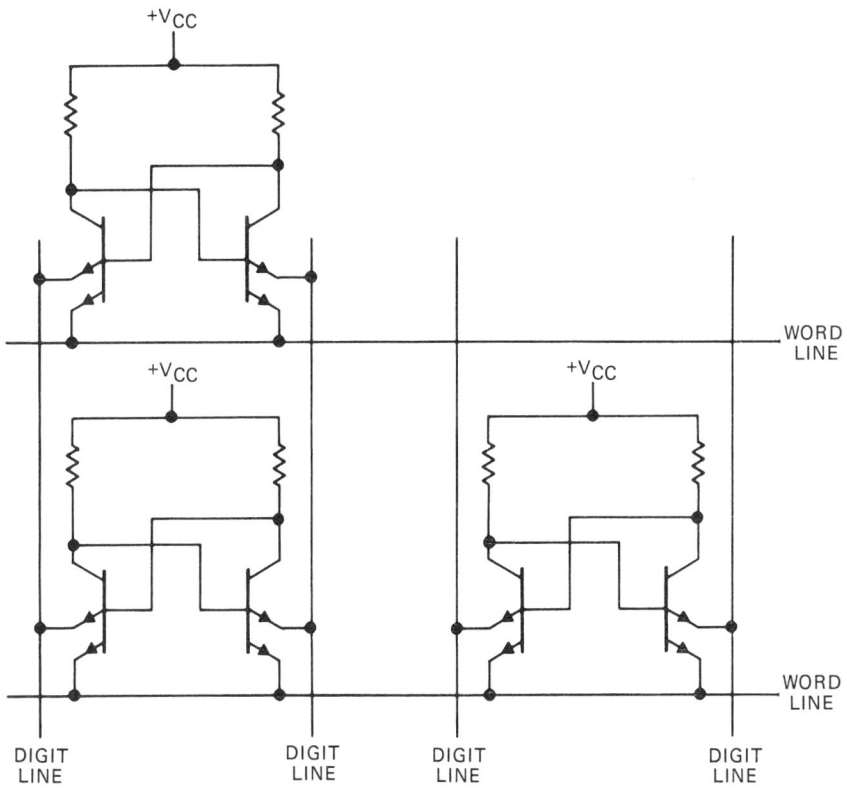

+V_CC

WORD
LINE

+V_CC   +V_CC

WORD
LINE

DIGIT   DIGIT   DIGIT   DIGIT
LINE   LINE   LINE   LINE

FIGURE 5-7

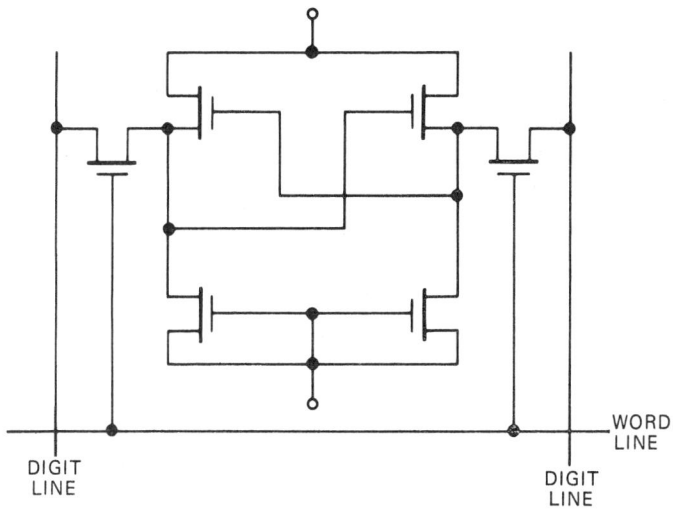

WORD
LINE

DIGIT   DIGIT
LINE   LINE

FIGURE 5-8

Figure 5-9 shows one form of MOS dynamic memory cell in which capacitor C is the memory storage cell. The difference between static cells (Figures 5-7 and 5-8) and dynamic cells is that information stored in a static cell remains fixed as long as power is maintained while in a dynamic cell storage is transient. In order to maintain information stored in a dynamic cell, it must be "refreshed" periodically. Dynamic memories employ charged capacitors as the memory elements. The charge leaks off and must be restored (refreshed). Leakage increases with temperature so that refreshing must be done more often at elevated temperatures. A typical refreshing at 70 C is carried out every two milliseconds.

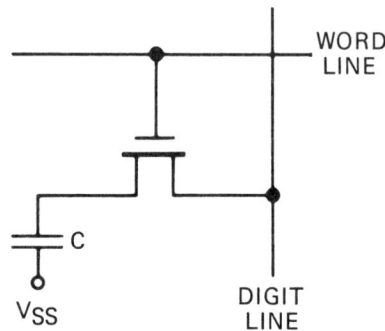

FIGURE 5-9

Writing into a dynamic memory comprises charging (or not charging) the storage capacitor C through the transistor connected to both word and digit lines. Reading is done by determining the charge on the capacitor. Reading may be done by means of a sense amplifier compatible with the low signal levels of the single transistor circuit. Although this single transistor memory cell requires more sophisticated auxiliary circuits, its small size is very attractive and it has been applied to 4K dynamic RAMs and some of higher capacity.

Figure 5-10 is another form of dynamic MOS memory cell comprising the gate capacitance of a MOS transistor C.

Figure 5-11 is still another form of dynamic MOS memory cell in the configuration of a flip-flop where the state is determined by charges on capacitors C_1 and C_2 (gate capacitances). Access is through two gate transistors.

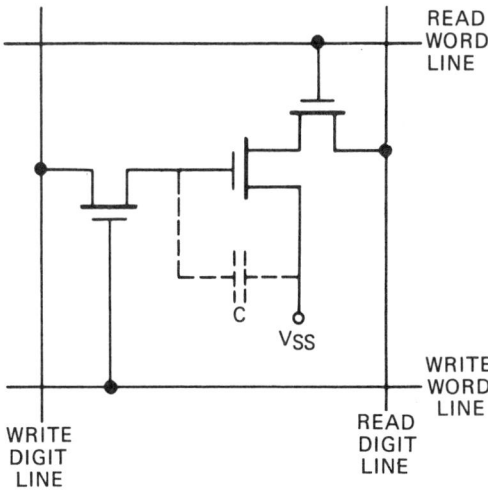

FIGURE 5-10

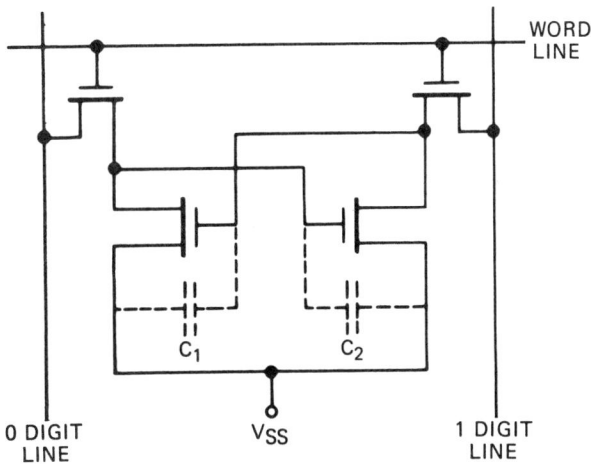

FIGURE 5-11

To generalize, bipolar RAMs are compatible with bipolar logic families, are faster, and consume more power than MOS. Developments continue, including Schottky, gold doping and better isolation providing substantial progress. Static MOS is slower than bipolar but smaller and less expensive. Low-threshold technology can be used to make them TTL compatible. The dynamic MOS comes the closest to competing with core memory systems in 4K units. They are smaller, faster, and consume less power than static MOS but require more exacting timing and other external circuitry. The choice cannot be made without weighing all factors. Generally speed and power consumption must be weighed against each other. Size, simplicity, cost and availability are some of the other important factors. Never forget, also, that availability from only one supplier may mean orphaned equipment at a later date. In a swiftly developing art, today's art may be radically changed tomorrow.

Permanent internal connections produce read-only-memories (ROM).

A "read only memory" (ROM) is a multi-element memory which has a meaningful pattern built into it. That is, it is made with a built-in coded message. The message is carried by circuits connected between predetermined x and y coordinates or access lines. The presence of a connecting link circuit means a 1 is stored and its absence means a 0. The circuits may be established by diodes or transistors.

Figure 5-12 shows a portion of a ROM formed of IC diodes. The diodes connected $X_2 - Y_1$, $X_1 - Y_1$ and $X_2 - Y_2$ represent stored 1's, while the open circuit to diode $X_1 - Y_2$ represents a stored 0. The mask used to provide the connections to the diodes contains the circuit completion or circuit open areas in accordance with the coded information pattern to be stored. Here we see the significance of "read-only" since once the pattern of interconnections has been printed on the IC chip, the stored information can be read but not altered. The basic cell simplicity of the ROM makes possible very high bit packing densities.

ROMs have many and growing applications. Wherever a fixed program is needed there is an actual or potential application for a ROM. Truth tables may be directly implemented without intermediate logic, providing means for implementing many computerized procedures. Character generation for cathode-ray tube displays use ROMs. Micro-programming makes use of ROMs. They are also used to store fixed user programs.

The pattern of connections or lack of connections providing the stored information is generally performed during the initial manufacture of the ROM. This is particularly true where large quantities of a

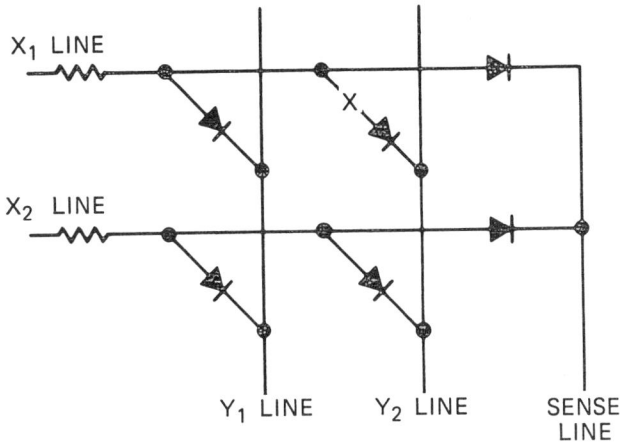

FIGURE 5-12

given pattern are required. Special masking can provide special patterns. However, cost may be a problem where only small quantities of a particular pattern are required.

To bridge the gap between ROM and RAM and to provide practical answers to the small quantity problem, the P/ROM has been devised—the P indicating "programmable." One way in which the programmable read-only memory (P/ROM) is made is to start with a basically bipolar pattern with all transistors cross-coupled to represent 1's at all intersections of x and y axes. Connections are made by means of fusible links. To change the initial 1's to 0's where desired, current is directed to selectively fuse the links producing the final pattern. This procedure permits stocking the basic ROM, permits the user to do the programming. This technique, which is only practical with bipolar logic due to the current levels required in the fusing process, is a big vote in favor of bipolar RAMs. MOS memories generally require customized masks which, although computerized, still add cost to the final packages.

Another method which has been used to provide electrically alterable RAMs employs a glassy semiconductor material electrically alterable between two stable states, amorphous or polycrystalline and having two degrees of conductivity. The alteration is carried out by means of controlled pulses.

Other methods of ROM alteration include a charged layer of silicon nitride between a metal gate and oxide insulator, a high voltage induced charge lasting for years, and an arrangement of floating gate P-MOS transistors were a high voltage between source and drain removes positive carriers from the floating gate, leaving isolated negative carriers on the gate creating a conducting channel. This latter technique provides a device which can be cleared by exposure to ultraviolet radiation which discharges the gate.

Multiplexers are formed of gates.

When the signals from several sources are to be carried over a single communications channel, they may be time multiplexed. Since this is basically a high speed switching problem, ICs provide a simple and effective implementation. All that is required is an array of gates and, if necessary, high speed logic gates where high switching rates are required.

Figure 5-13 shows a 4-to-1 multiplexing circuit. Which of input

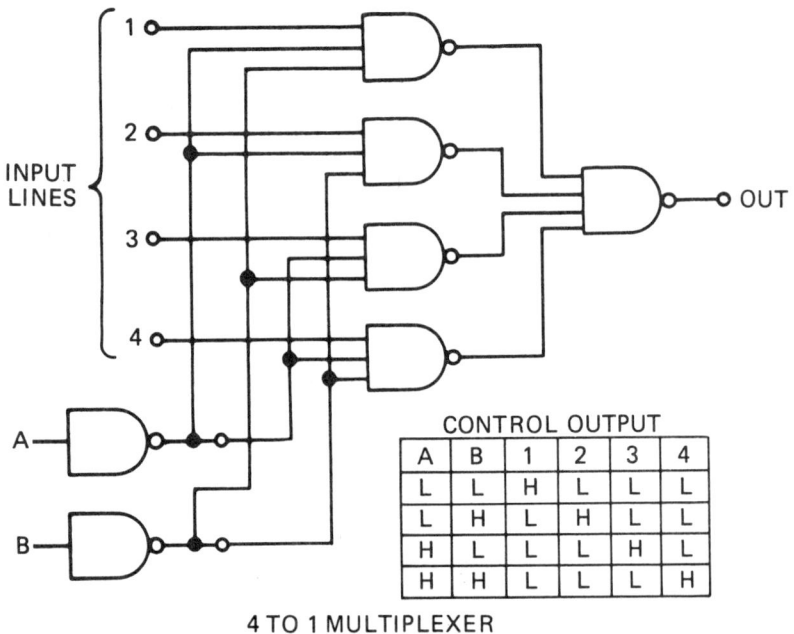

CONTROL		OUTPUT			
A	B	1	2	3	4
L	L	H	L	L	L
L	H	L	H	L	L
H	L	L	L	H	L
H	H	L	L	L	H

4 TO 1 MULTIPLEXER

FIGURE 5-13

lines 1 through 4 is to be outputted is determined by the states of
control lines A and B. The truth table shows the high/low conditions of
lines A and B to provide high output put-through. Thus, when A and B
are low, the output is high for input line 1; when A is low and B is
high, the output is derived from input line 2, and so on.

Encoders and decoders are also formed of gates.

The philosophy of the encoder and decoder is much the same as
that of the multiplexer. Specifically these most useful ICs are designed
to convert a given input pattern to a different but predetermined output
pattern. An encoder, for example, may be designed to receive binary
coded input signals and output Gray code signals. Figure 5-14 shows
one simple circuit for converting binary coded signals to Gray code
coded signals. It makes use of three exclusive -OR gates. This could be
called an encoder or a decoder, depending on your point of reference.

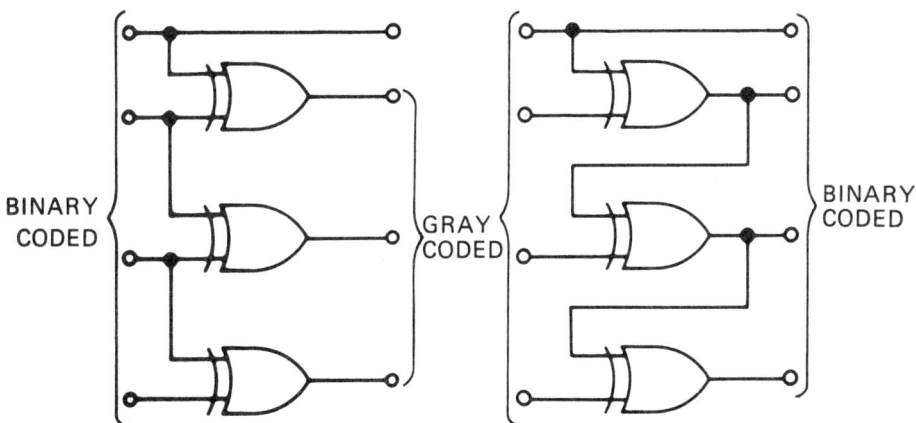

FIGURE 5-14

Figure 5-14 also shows the inverse, i.e., an encoder or decoder, as the
case may be, for converting Gray code coded signals to binary code
coded signals. Again this is accomplished with three exclusive -OR
gates.

Thus, encoding and decoding are rather straightforward uses of
combinations of gates. The design involves a statement of the input
code and the desired output code. Truth tables are very helpful here,

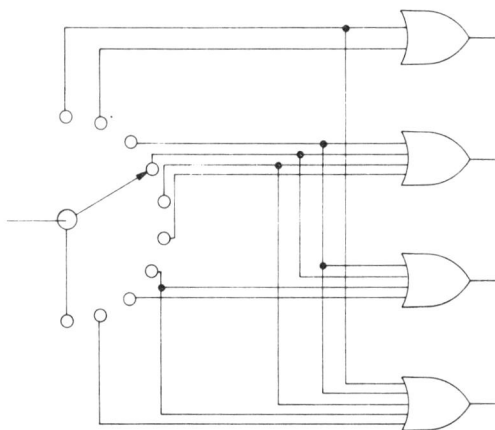

FIGURE 5-15

not only in setting up the original proposition but also in stating how it can be accomplished. Due to the fact that NAND and NOR gates are less expensive, requiring fewer transistors per gate, most encoders and decoders use these types. Due to the almost universal need for many specific encoding and decoding functions, a substantial number of often used configurations are available in MSI packages.

Code converters are very useful combinations of gates, too.

Since, as stated above, encoding and decoding, as well as some other useful conversions such as readout lamp drivers, all involve the analogous process of encoding and decoding, such devices will be termed code converters. First, the most obvious conversion is decimal to binary (as shown in Figure 5-15), where the decimal input is represented by a ten-point selector switch with contacts representing digital numbers 0 to 9 and the output lines provide 1, 2, 4 and 8 binary values. This process can be extended to represent decimal numbers higher than 9 but the number of inputs to the various gates rapidly increases to a point where conversion becomes impractical. One of the most popular of the practical forms of decimal to binary conversion is called binary coded decimal (BCD). In this code each digit of the decimal number is individually converted to a binary number. The converter above is typical of a converter for one decimal digit of a BCD converter.

Conversion from binary to decimal is also a very important process. The gating diagram of Figure 5-16 is a typical format for converting BCD (1248) to decimal 0 to 9. For purposes of illustration and demonstration, the circuit shown is made up of four inverters and ten AND gates. Such devices are readily integrated on a tiny chip and packaged in a typical 16 pin DIP. This is a good example of what integration can provide in the way of a multiple-element building block. A commercial form of BCD-to-decimal decoder uses the equivalent of four transistors in each of the output gates and two in the input gates, making a total of 48 transistors in all. The evolution from 48

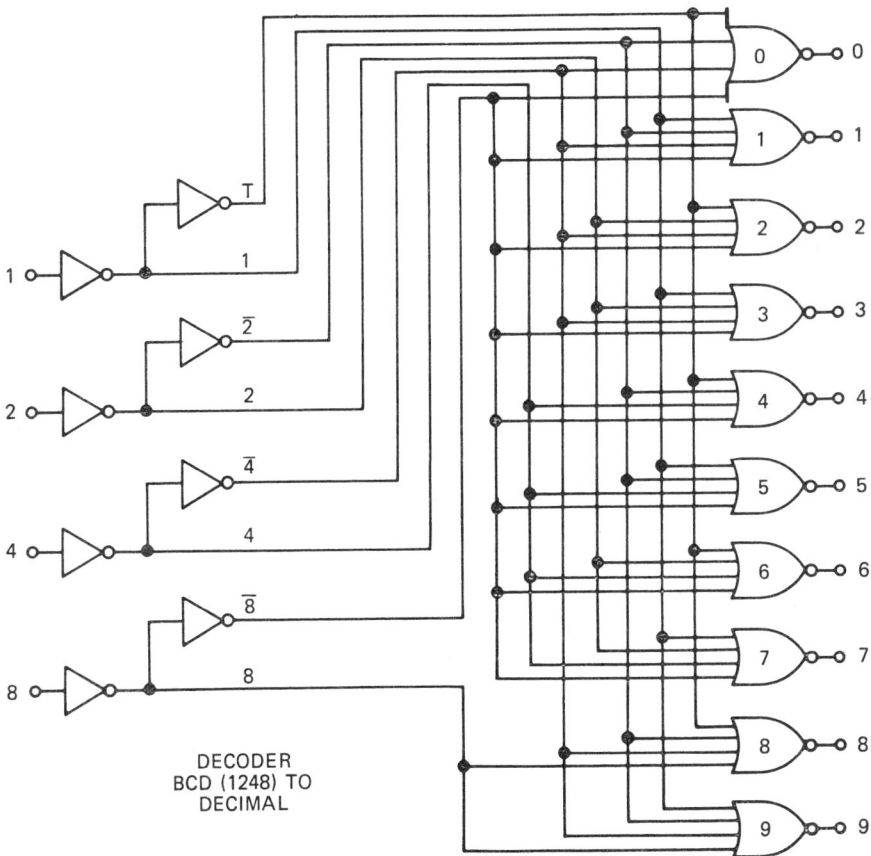

FIGURE 5-16

vacuum tubes to 48 transistors to one tiny package in the space of 25 years highlights some of the amazing progress in the field of electronics which has taken place over the years.

The data selector is a useful gating concept.

Data selector is a name applied to a circuit very much like a multiplexer (described earlier). A 4-channel data selector provides means whereby any one of four input lines can be connected to a common output in response to a predetermined combination of binary coded "select input" signals applied to data select terminals.

Adders are basic to logic computers.

The most basic step in performing arithmetic by ICs is addition. The step is performed independently on each binary bit of the addend and augend. The result of adding two binary numbers is their sum and carry. The binary sums are as follows: $0+0=0$; $0+1=1$; $1+1=0$ with 1 to carry. There are two commonly recognized adding circuits; one is called a half-adder and the other a full adder.

Figure 5-17 an exclusive -OR circuit and AND gate which together form a so-called half-adder. That is, they provide the sum output $A+B$ and the carry output AB. In other words, if A and B are both 0, the sum is 0; if either A or B is 1, the sum is 1; however, if A and B are both 1, their sum is 0 with a carry 1. This is shown by the truth table forming a part of Figure 5-17.

It will be noted that the circuit of Figure 5-17 has no provision for

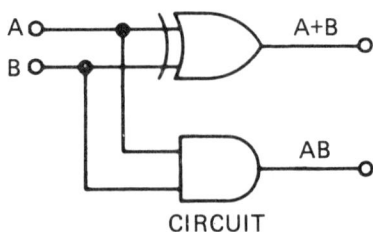

A	B	SUM	CARRY
0	0	0	0
0	1	1	0
1	0	1	0
1	1	0	1

TRUTH TABLE

HALF ADDER

FIGURE 5-17

accepting three inputs, i.e., A, B and carry. This is the distinction, since the full-adder will accept a carry input. The full-adder is formed by cascading two half-adders as shown in Figure 5-18.

The full-adder of Figure 5-18 will accept two inputs to be added, A and B, and a third input, carry, which also must be added. The resulting truth is shown as a part of Figure 5-18. This circuit then will perform the complete addition function, including a possible carry-in from a lower significant digit.

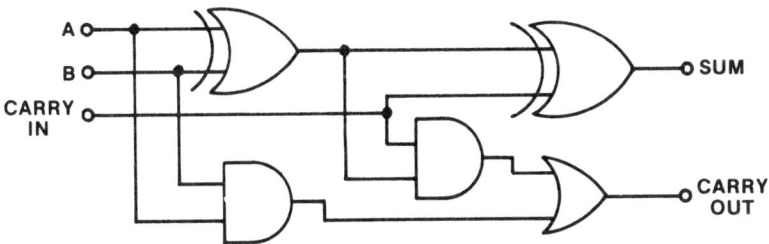

FULL ADDER

INPUT			OUTPUT	
A	B	Ci	S	Co
0	0	0	0	0
0	0	1	1	0
0	1	0	1	0
0	1	1	0	1
1	0	0	1	0
1	0	1	0	1
1	1	0	0	1
1	1	1	1	1

FIGURE 5-18

Other arithmetic is performed by modifying the adding function. Subtraction consists in first, complementing the subtrahend, adding this to the minuend, and adding 1 to the result. Multiplication consists in multiple addition, once for the number in the multiplier. Division consists in the inverse of multiplication. These arithemetic processes are performed dynamically a digit at a time. The numbers being manipulated and partial results are held in registers from which they are extracted and to which partial results are returned between operations.

Analog-to-digital converters are an interface between the real world and the world of ICs.

Most sources of electrical signals in the real world provide analog voltages or currents. These are not compatible with computer processing and even present problems in transmission over communications links. Thus conversion of these analog signals to digital form is of vital importance in a computer oriented world. There are many ways in which analog-to-digital conversion can be implemented. The choice of which method to use depends on a number of factors, including speed, accuracy, resolution, linearity, stability, input ranges, digital output codes, physical size and cost.

Cost is a factor since analog-to-digital converters are available over a wide price range, and choosing a converter of greater capability than is required for a given application could result in paying a substantial, unnecessary premium. A knowledge of the problem requirements and device capabilities will be largely responsible for choosing the right converter in the $50 to $5000 price range.

Other choices may also be important. For example, where a number of signal sources are to be converted from analog to digital, should they be multiplexed or should a separate converter be provided for each analog source? Again, if multiplexing is indicated, should it be carried out at the analog level or after conversion to digital? A one-shot event such as an explosion cannot be multiplexed since many results must be monitored simultaneously. Also, very fast-changing signals, such as those produced by vibration of a plane in a wind-tunnel test, cannot be multiplexed or vital data will be lost. The most likely candidates for multiplexing are slowly changing signals such as those derived from a thermocouple.

Multiplexing before the analog-to-digital conversion is the most economical since A/D converters, even today, can be classed as relatively expensive components. However, where signals are of very low level, multiplexing may introduce objectionable noise. One solution here might be to amplify the source signal before multiplexing.

There is no such thing as a direct, simple conversion of an analog signal to its equivalent digital signal. The process consists in compar-

ing a reference signal having a known digital value with the analog
signal; of changing the reference signal (having a tracked digital
known value) until a match is detected between it and the analog
signal; and then outputting the known digital signal as the equivalent of
the analog signal converted to digital. The details of how this process is
carried out provide the different forms of analog-to-digital converters.

Perhaps the easiest to understand is ramp comparison, shown in
much simplified form in Figure 5-19. A timed ramp voltage is generated and compared with the analog voltage to be converted in a comparator. The ramp voltage rising at a known rate is timed by a clock
pulse generator so that at any instant the number of clock pulses generated since the start of the ramp voltage is a measure of that instantane-

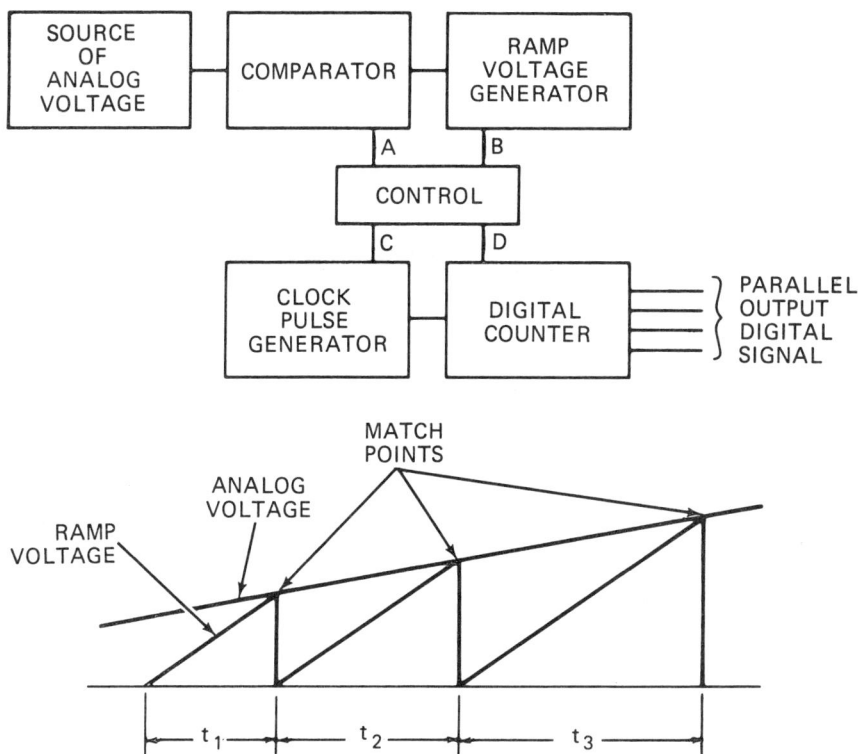

FIGURE 5-19

ous voltage. A digital counter converts these pulses to a parallel output digital number representing the instantaneous voltage. Now, if the clock and ramp voltage are started from zero at the same instant, and a comparator compares the instantaneous ramp voltage with the analog voltage to be converted, when a match is sensed the control stops the clock and reads out the contents of the digital counter. This output is then the digital equivalent of the analog source voltage. The ramp and clock are reset to zero and the process is repeated.

It will be seen that what has been done is to generate an analog voltage having a known digital value; to compare this analog voltage with the analog voltage to be converted; and when equality is detected, to read out the known digital value. There are many ways in which this basic idea can be implemented. As digital circuitry becomes more and more adaptable and better understood the trend is toward more purely digital techniques.

One well-known A/D converter is shown in block form in Figure 5-20. Here a counter driven by a clock generates a count, which when converted by the digital-to-analog converter, provides a staircase ramp

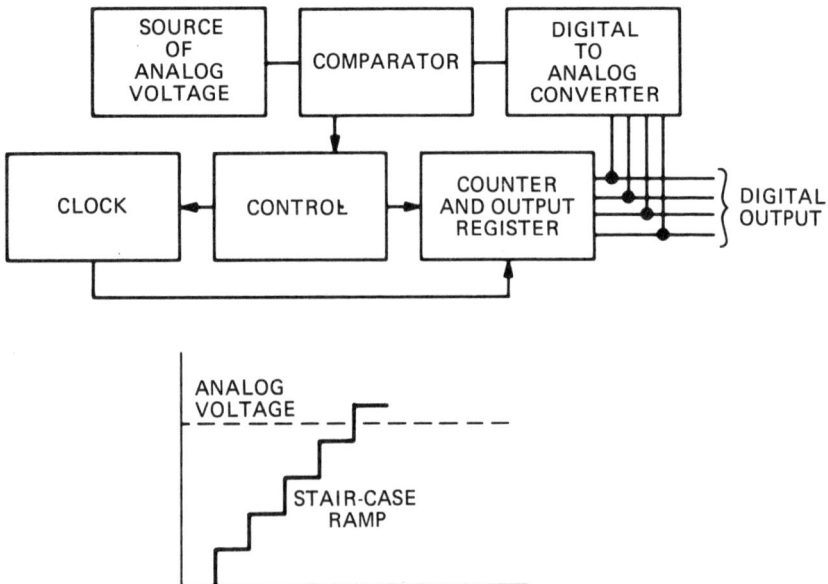

FIGURE 5-20

voltage which in turn is compared with the source of analog voltage to be converted in the comparator. When the ramp voltage passes through the analog voltage level, the comparator signals the control which stops the clock and reads out the count in the counter through its output register. The counter is reset to zero and the clock is restarted to repeat the process. It will be noticed that the match between the staircase ramp and the analog voltage is not perfect since the reference voltage occurs in steps. If each step is equal to the last significant bit (LSB) in the digital readout, the output has an ambiguity of $\pm$ 1/2 LSB.

If the comparator, instead of having a single input has several inputs, each provided with a predetermined fraction of the reference voltage as its reference, a parallel A/D converter can be provided which reads three or more bits simultaneously.

While the A/D converters described above are unipolar devices, bipolar converters can be provided with a few modifications. The reference, for example, is made bipolar. The comparator accepts positive or negative inputs. The digital output must show a polarity sign.

Signal conditioning for A/D converters can improve accuracy.

The A/D converter standing alone has a number of limitations in the real world which make auxiliary devices very often necessary. These auxiliary devices are generally required to make the analog signals to be converted compatible with the A/D converter or to improve their characteristics. These devices can generally be classified as signal conditioning means.

Preamplification is perhaps the most common conditioning device. For example, the millivolt signals from thermocouples must be amplified to a level compatible with the A/D converter intput, which is typically 5 or 10 volts full scale. This preamplifier will generally be an operational amplifier having a predetermined gain. It must have a slew-rate and linearity compatible with the signals to be converted. If an extremely wide range of signal levels is to be covered, the preamp function may call for a logarithmic response amplifier. If there is a common-mode voltage problem, the preamp must be able to cope with it without introducing an intolerable error.

Filtering is another common signal-conditioning device. If the signal being converted has a slow enough rate of change, filtering can be very effective in removing noise components from the signal. If simple, brute force filtering is impractical, signal averaging may sometimes be resorted to in order to extract the signal from the noise.

In cases where the signal may change during the finite A/D conversion period, the sample-and-hold technique is often used to prevent ambiguous results. The sample-and-hold device samples the analog signal periodically but holds a steady sampled value while the A/D converter is performing its function. (Sample-and-hold devices, functions and applications are set forth above.)

Isolation to eliminate noise pick-up or grounding problems may be resorted to. By isolation is meant that a signal is separated from any ground connection so that it is not modified at a distant point by a different ground reference potential. As an analog signal, isolation can be accomplished by modulating a suitable carrier, transmitting the modulated carrier to a distant point as an ac signal and then demodulating it to recover the initial signal. Isolation after conversion to digital form of signal is most effective. Digital coupling can be performed with optical isolators providing a very high degree of isolation, and low impedance digital signal circuits are highly noise resistant.

When filtering or otherwise conditioning a signal for further use, it is important to keep in mind the important aspects of the signal to make sure they are not sacrificed. If sudden abrupt changes in signal level are important, the A/D converter and all of its signal-conditioning accessories must respond to the abrupt change being sought. This means that filtering should not obscure it and the A/D resolution should be fine enough to process it.

The following factors are important in specifying an A/D converter:

Characteristic	*Range*
Resolution Bits	8-10-12-16-17
Conversion Time	1 us; 1-10 us; 10-50 us; 50 us
Range (Input)	Fixed, selectable
Output	Serial, parallel
Codes Available	BIN; BCD; other
Temperature Coefficient	10-25-50 ppm/°C
Packaging	Card-Module
Cost	

While these are the basics, there are some fine points which pertain to particular models and particular manufacturers. Linearity, for example, is important in some applications not covered above.

Digital-to-analog conversion back to the real world.

The digital-to-analog (D/A) conversion process is performed open loop and is much easier and simpler than the A/D process described above. The process starts with a digital signal which has a definite and predetermined connotation.

The most common and basic method of D/A conversion employs a source of reference voltage or current, a series of binary-weighted resistors, a set of switches and an operational amplifier. Figure 5-21 is a simplified diagram of a typical D/A converter including these basic elements. The switches are indicated as being closed in response to the presence of bits in the digital number applied from the digital input. This input can be taken to represent any suitable source of digital binary coded information to be converted to an analog voltage having a full scale value of EREF. The output voltage Eo is equal to (EREF R/2)/R′ where R′ is the equivalent resistance of all of the resistors connected in parallel by digital activated switch closures. Thus, with the bit switch closed, Eo = EREF R/2 = EREF/2 and with all bit switches closed, Eo = EREF − LSB = EREF (1 − 1/16) or Eo = 15/16 EREF.

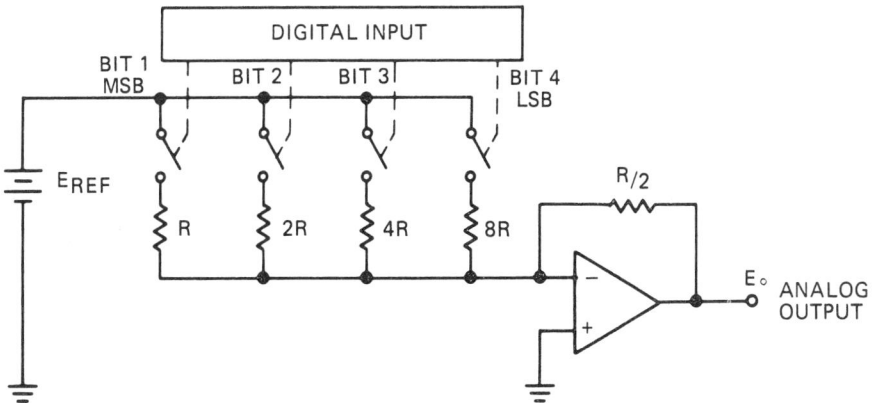

FIGURE 5-21

While the basic concept of D/A conversion is simple and straightforward, there are many practical problems, and when used in a system, many complications can arise. One of the first problems concerns the resistors required for the conversion. In the example above, if

R is 10K, 8R will be 80K. It is generally desirable to provide these resistors in integrated form, either thin film or thick film. While the above values may be integrated, problems arise when the converter is carried to more bits. For example, an 8-bit converter would require a resistor of 128R or 640K and would present a real problem in integration. One way of reducing the range of resistors required is to group them with attenuation between groups. Figure 5-22 is a simplified diagram of a D/A converter accepting 8 bits of digital input. The largest resistor required is 8R, since after the first group of resistors starting with the most significant bit (MSB) there is a series resistor equal to 8R, which in effect divides the contribution of each of the four bits in the following group by 16.

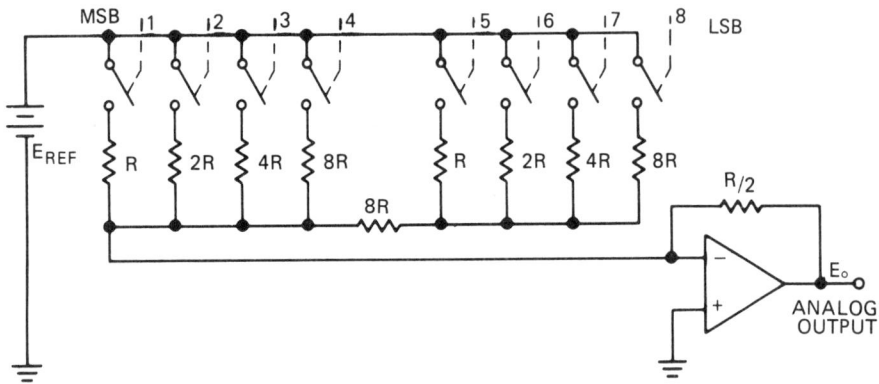

FIGURE 5-22

A still further improvement may be obtained in a ladder network as shown in Figure 5-23. The output depends on the number of bits receiving logic 1 inputs as set forth above, i.e., the first or MSB provides $E_{REF}/2$, the next bit $E_{REF}/4$ and so on, the total output being the sum of ON bits. The great advantage of this last circuit is that integration of only two values of resistance is required and these can be nominal values of, say, 5,000 and 10,000 ohms. One resistor can be eliminated by a unity gain connection of the amplifier using the same resistor network as shown in Figure 5-24. Weighted current sources may be substituted for the reference voltage source, providing still another basic configuration possible for a D/A converter.

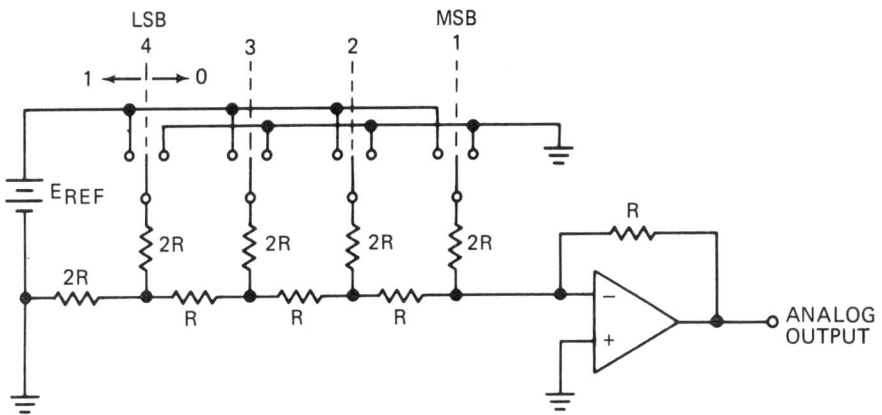

FIGURE 5-23

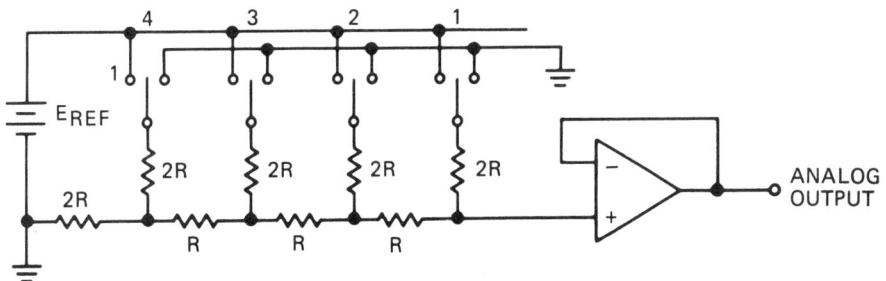

FIGURE 5-24

While a battery symbol has been shown representing the reference voltage source, in practice batteries would not be stable enough to provide accurate conversion. The most common reference voltage source is a temperature-compensated zener diode provided with a constant current. The current should be supplied in such a manner as to maintain the current through the zener diode constant regardless of the current drawn by the switched resistors in the converter network.

In case the required output is bipolar as represented by a digital offset binary or 2's complement code, an offset reference can be added to provide the bipolar response. This may be done by merely supplying a negative reference current equal to one-half the maximum positive reference current and increasing the amplifier gain by a factor of 2 as shown in Figure 5-25. Thus, zero output will occur when the positive reference current reaches one-half of its maximum value. The doubled gain is necessary since the output excursion is doubled from 0 to 10 volts to 0 to ± 10 volts. A sign-magnitude code can be implemented by means of two additional operational amplifiers, one of which is used for positive output polarity voltage and the other for negative. The MSB is used to operate a polarity switch which turns on the appropriate polarity amplifier.

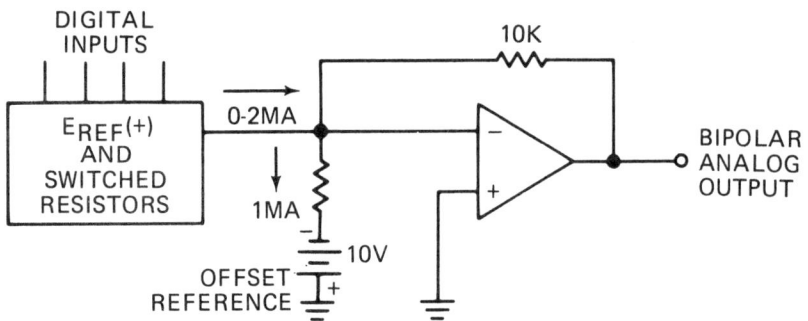

FIGURE 5-25

The accuracy with which a D/A converter operates depends on all of its component parts. Operational amplifiers are available with a wide range of characteristics, of which offset stability is the most important characteristic in the D/A context. The offset and its drift with time and temperature, as seen at the amplifier output terminal, should be less than the equivalent of the least significant bit (LSB) being converted, also as seen at the output terminal. Of course, the amplifier must be capable of operating linearly to the full maximum required analog output, generally ± 10 volts, and into whatever load is to be driven. Where conversion of 12-bit accuracy is required, requiring overall error of no more than about 0.01 percent, trimming will generally be required for overall gain and offset bias. How this may best be accomplished will depend on the actual circuits used in the converter.

Since the absolute value of D/A converted voltage is no more accurate than the accuracy of the reference voltage, a high stability temperature-compensated zener supplied with constant current is indicated. It is noted that the reference voltage called for is generally 10 volts while the ideal zener voltage is 6.3 volts. This means that the reference must include a voltage repeater which can be adjusted to have an accurate gain of 10/6.3 to supply the 10 volts from the 6.3 volt source. Such a repeater must employ accurate, low-temperature coefficient resistors and a stable operational amplifier.

The programmed resistor network enters directly into the conversion process. However, if the ratios of the resistor values are precise, their absolute values are of secondary importance. This fact has two important implications as far as integrated circuits are concerned. First, integration of resistor networks having constant ratio values is much easier than integration to provide absolute values. Second, the temperature coefficient of a resistor network is substantially the same for all of its resistors so that ratios are preserved even if absolute values change. Thus, the converter resistor network is a product readily integrated to provide a commercially useful, low cost product.

The final component of the basic D/A converter is the operational amplifier—again a well-known product available in a wide range of characteristics. It is merely necessary to choose the best compromise between performance and cost. Offset stability is again a primary consideration. The basic circuit calls for an operational amplifier, typically one which can provide ± 10 volts output within its linear range. However, the analog voltage produced by the conversion is presumed to have some practical application. The particular application may require only a nominal current from the amplifier as, for example, in driving a graphic recorder, or it may require some real power, as in automatic testing systems. In the first instance the average operational amplifier would be quite adequate to provide the output drive. In the second instance the power required by the test system load would require a power stage after the first operational amplifier. Operational power supplies are also available for supplying a wide range of power, and some incorporate the D/A conversion function providing a unitary power converter.

6

A Key to Effective Circuit Design Is Breadboarding

Defining breadboarding.

Originally breadboarding meant just that: the building of an electronic circuit on a breadboard. The purpose was to try out a circuit, experiment with it and refine it before committing it to final form. The breadboard was a natural. It was an insulator so that circuits could be mounted directly on it. It was inexpensive, a boon to early experimenters. A circuit could be spread out over the board, physically following the schematic diagram and thus providing a visual bridge between the schematic and the actual circuit. It was ideal for change and experimentation since parts were easily moved around and nothing was permanently fixed in place.

Why one should do breadboarding.

Breadboarding is generally useful when temporary, modifiable circuitry is needed. It allows trying out new circuits without commitment to a final form. It allows experimentation to determine the most efficient way to carry out an idea. It allows circuit analysis to determine limits of operation; the effect of temperature and supply voltage; the effects of aging; temperature rise and hot spots; and to analyze circuit malfunction. Circuits may be opened to measure currents and impedances at various points. Test points are readily accessible. The effects of lead dressing and noise pick-up can be determined. Worst-case conditions can be simulated and adverse effects countered. Components are not damaged and so can be used over and over again.

Many useful devices are available.

Many different types of breadboarding aids are available particularly directed to integrated circuit use. One of the simplest and most basic is a small board with spring contacts as shown in Figure 6-1. The contacts are spaced so that DIP mounted ICs can be combined with other DIPs and discrete components. Such a board can be used to set up and finalize simple circuits and can be used over and over again.

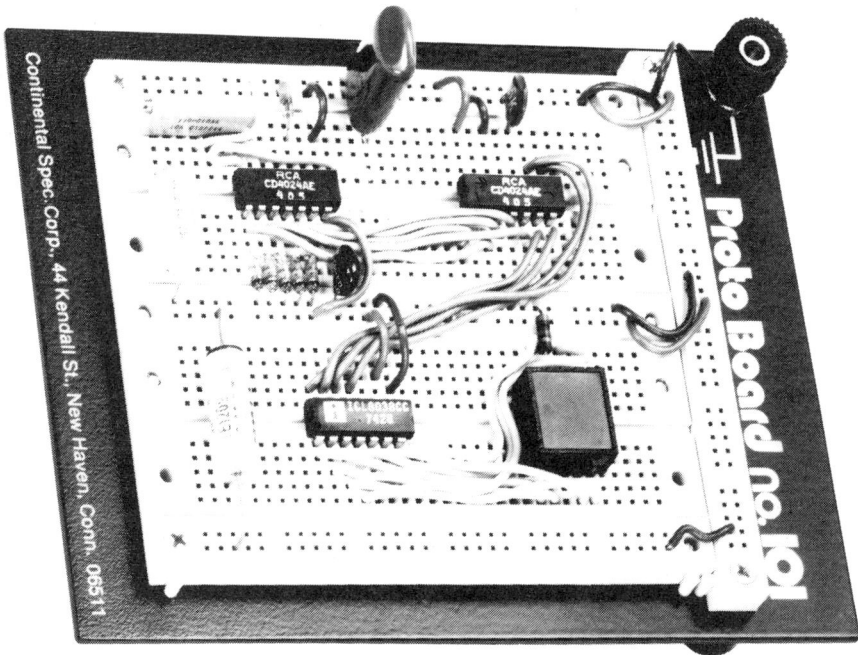

FIGURE 6-1 (Courtesy Continental Spec. Corp.)

One very useful arrangement I have made and use continually consists of two DIP sockets mounted on a Bakelite box with numbered pin tip jacks along the edges as shown in Figure 6-2. Two of these boxes (four sockets) permit combinations of up to four ICs. A five-

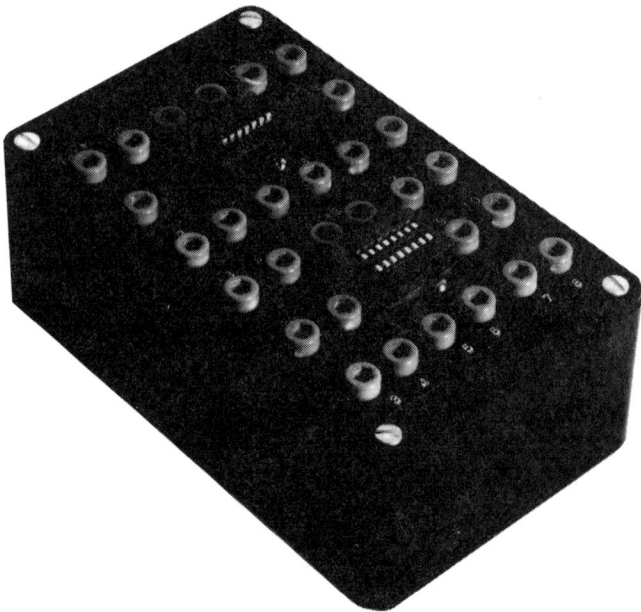

FIGURE 6-2
Experimental Circuit Box Providing Two DIP Device Sockets

push-button input box with two LED output indicators is a companion unit as shown in Figure 6-3, and the internal connections are shown in Figure 6-4. The push-button box was laid out and labeled particularly for experiments with J-K flip-flops. However, I have found it useful in making many other experiments, for example, the one described below.

Describing an experiment to develop an EXCLUSIVE -OR circuit.

In Chapter 2 (Figure 2-19), an exclusive -OR gate was shown and described. Many circuits have been proposed and used to execute the exclusive -OR function. Some using discrete components are very simple. Some have been integrated which seem unusually complex. The point here was that I wanted an exclusive -OR gate using simple gate modules. When I duplicated some of the published circuits I found

FIGURE 6-3
Experimental Circuit Box with Push Buttons and LED Indicators

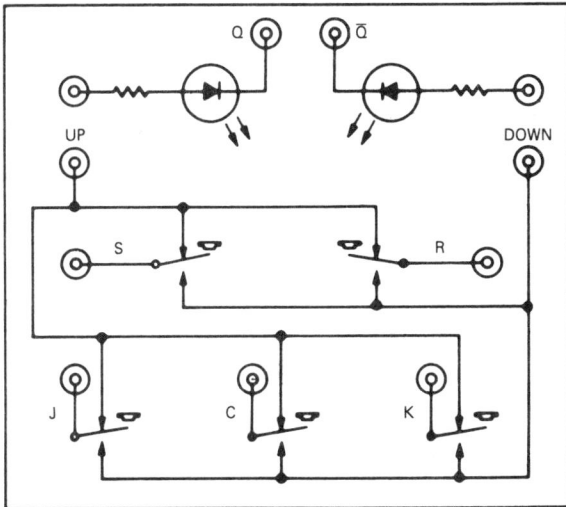

FIGURE 6-4

it necessary to use two or three DIP packages. Using the socket boxes described above I finally arrived at an exclusive -OR circuit which could be made using a single 7400 octal 2- input NAND gate DIP. The experimental set-up is shown in Figure 6-5. The J and K push-buttons and one of the LED indicators of the J-K box (Figures 6-3 and 6-4) were used in this experiment.

FIGURE 6-5
Exclusive-OR Gate Using Experimental Circuit Boxes

Breadboard laboratories provide very complete facilities.

Special printed circuit boards, with patterns to accommodate ICs but without interconnecting wiring, and with edge connectors, are available for a somewhat more advanced form of breadboarding. Since these boards provide for soldering the ICs in place they are far less flexible than the forms of breadboarding described above. They are useful where one is rather confident of the workability of the circuit but wants to try it out in as close to final form as possible before ordering special printed circuit boards. They permit checking out the circuit and a certain amount of experimentation but may result in damaged components if extensive changes have to be made.

Breadboard laboratory.

Several systems or set-ups are available which provide for the breadboarding of substantial systems. One of these is the Elite 2 shown in Figure 6-6. These are relatively expensive and may include such aids as power supplies, signal sources and logic state indicators. They are useful for teaching, experimenting and setting up more or less complete control or other integrated circuit system. Some of these are designed specifically for teaching and demonstration while others are more directed to large scale breadboarding of new systems.

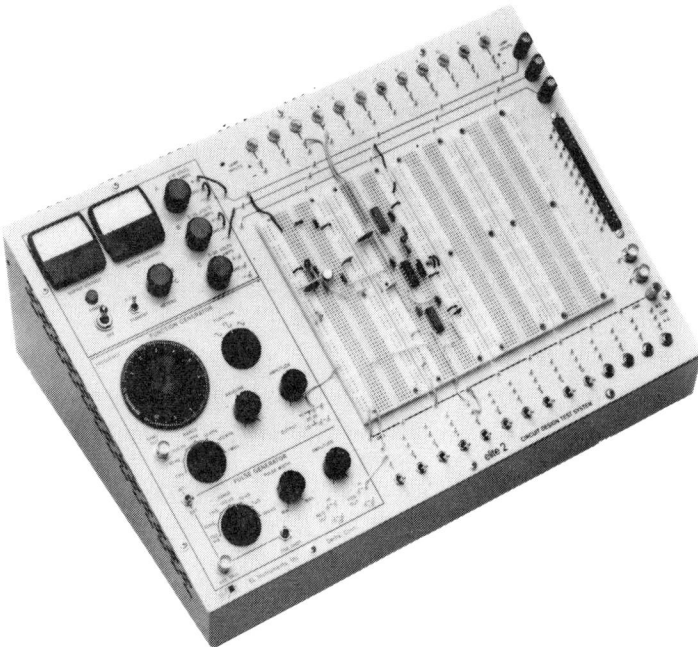

FIGURE 6-6 (E&L Instruments, Inc.)

Simple but interesting circuits are easily set up with experimenter's kits.

Under the general heading of breadboarding should be included certain types of experimenter's kits. These are the kits which contain sufficient components to construct a large number of experimental projects, one at a time. Experimenting can be an interesting hobby and it can also be a way of learning more about a given electronic technology. With a relatively few basic modules and auxiliary parts, a large number of different devices can be built, tested and modified. One of the problems with most experimenter's kits, however, is that the accompanying instructions are often meager and even misleading. An experimenter should not only be interested in making an experiment work, he should be even more interested in understanding how it works and want to be able to innovate on his own for added effects. Instructions which enable one barely to make a circuit operate but without any real understanding accomplish little as far as the learning process goes. Add to this, non-standard symbology and even poor English and a kit is more a disaster than a help.

One needs several auxiliary devices to breadboard properly.

In addition to the actual sockets and interconnecting means of a breadboard, certain auxiliary equipment is necessary. First of all one needs one or more power supplies. A convenient and useful form of power supply is one with an adjustable and metered output voltage to cover the IC range of 3.75 to about 15 volts. It should have internal short-circuit protection and preferably adjustable current limiting. An output current meter is also useful but not absolutely essential.

Simple gates respond to any change in level traversing their input switching points, whether dc level, pulse or repeated pattern. Many more complex devices such as flip-flops, counters, registers and so on may respond only to specific forms of input. Simple dc level changes can be provided by means of a potentiometer bridged between two voltages embracing the 0 and 1 input levels required. Fast and clean level changes can be generated by means of a Schmitt trigger as shown in Figure 6-7. The capacitor across the input filters out rough contact closures in the input switching. The output is inverted and shows very fast change from 0 to 1 or 1 to 0, as the case may be. A very versatile pulse generator (555) having a number of useful applications is described below.

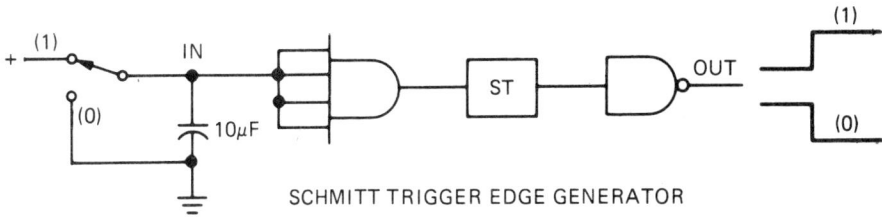

SCHMITT TRIGGER EDGE GENERATOR

FIGURE 6-7

Output or logic state indicators are also needed. While a dc voltmeter will give precise information as to input and output levels and the margin between the two and hence should be at hand, simple on-off indicators provide much necessary information, especially in a complex circuit. Light-emitting diodes (LED) with suitable current-limiting series resistors are ideal logic state indicators. Use them generously.

How to solve some breadboarding problems.

Ripple, noise and common impedance of the bias voltage source are less troublesome in digital circuit systems than in analog since the digital circuits have a threshold of response below which extraneous signals have no effect. Analog circuits are, on the other hand, affected adversely by the slightest amount of unwanted signal voltage. In addition, the analog circuits themselves add noise and distortion which are cumulative. Digital circuits purge themselves with each stage.

The factors most likely to cause problems in digital circuits are improper bias voltages, poor socket contacts, bent pins of plug-in devices, poorly soldered joints and wrong connections. All soldered connections should be visually inspected, preferably with a low power magnifying glass, to see that all soldered connections have been properly made, i.e., that the solder has "wet" both parts being joined and that no rosin separates the solder or the parts. Also look for solder bridges between parts or contacts not to be joined.

Make sure you have properly identified device terminals and leads. Don't take it for granted which side is intended to be up in the picture. Check and recheck all connections before applying power. When first applying power, watch to make sure there isn't an improper connection which will be indicated by excessive current being drawn from the power source. Make sure, when excessive current is indicated, that the bias voltage source is within specified limits for the particular circuit being powered. Also check to see that the IC package is plugged in with the proper orientation and that the bias voltage is of the correct polarity. A typical package of TTL gates drew 7.5 ma. no load when properly connected and 215 ma when either package or bias source was reversed.

A regulated power supply with an adjustable current limit control is almost a "must" in breadboarding experiments since it can help avoid costly results due to improper connections. When such a power supply has an indicator of over-current conditions, it is even more helpful in preventing damage to components.

A versatile timing circuit that can be very useful.

A monolithic timing circuit, suffix -555, manufactured by several companies, is a versatile device capable of generating pulses, delays or oscillation. Operated as a monostable multivibrator, an external capacitor and resistor are used for the timing network (see Figure 6-8).

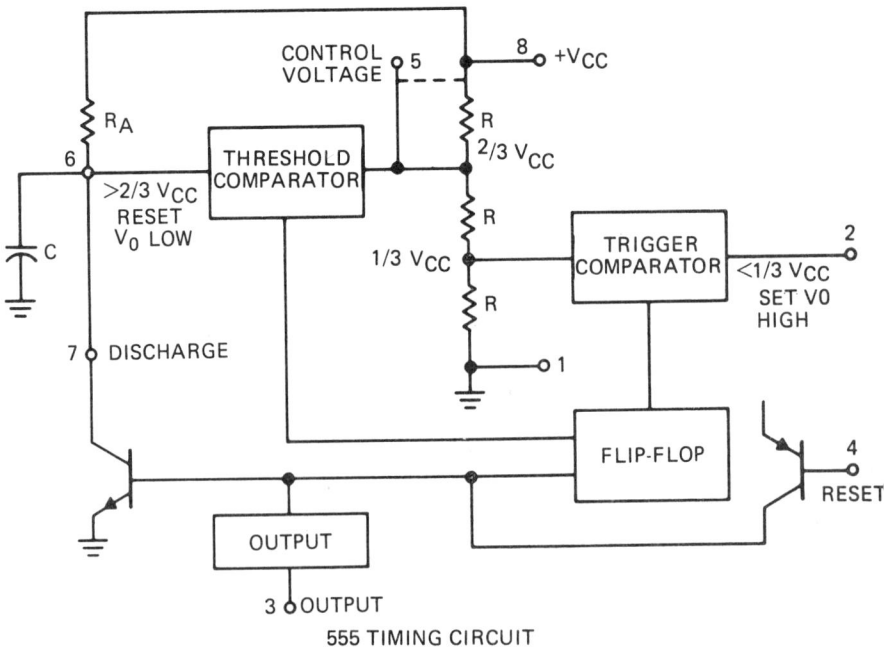

555 TIMING CIRCUIT

FIGURE 6-8

Internally the integrated circuit -555 provides two comparators, one for the input signal and the other for the capacitor voltage, and a flip-flop and output driver. Since the comparator reference voltage is chosen as a fixed ratio of the Vcc voltage, the timing is independent of the supply voltage.

A simple change converts the -555 into an oscillating astable circuit. This is done by merely connecting the capacitor back to the trigger input as shown in Figure 6-9.

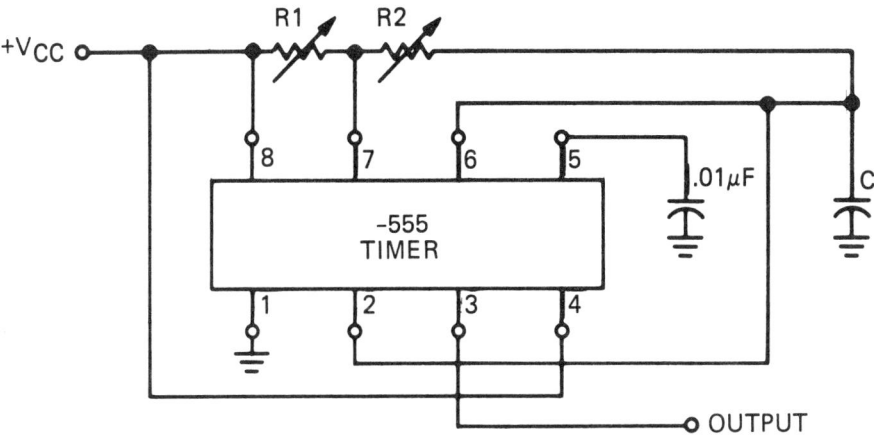

FIGURE 6-9

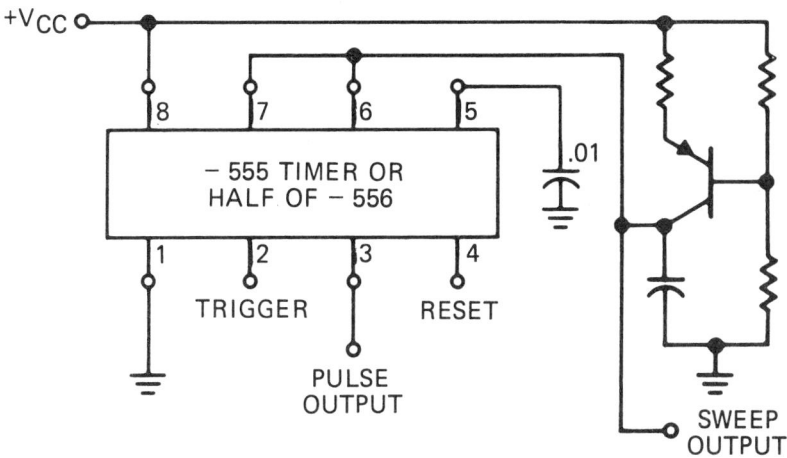

FIGURE 6-10

Another simple change converts the -555 into a linear voltage sweep generator or linear voltage ramp. This is accomplished by providing a constant current transistor circuit to charge the external timing capacitor and taking the output voltage across the same capacitor, as shown in Figure 6-10. The driving and resulting wave forms are shown in Figure 6-11.

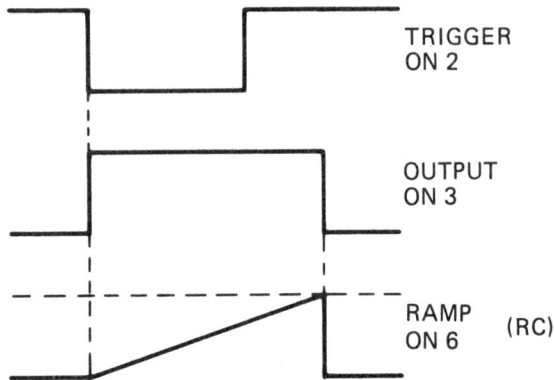

TRIGGER
ON 2

OUTPUT
ON 3

RAMP
ON 6   (RC)

FIGURE 6-11

The -555 has also been adapted to operate as a missing pulse detector, a pulse width modulator and a sequence tester. This most versatile device should be made a familiar tool in any integrated circuit test or experimental set-up.

Additional points concerning breadboarding.

Circuit test points, i.e., points of significance to circuit operation, are always useful in permanent circuit boards as well as in breadboards. These can generally be provided at little added cost or complexity and without affecting circuit operation.

Breadboarding is seriously at a disadvantage in one important area, that is, at very high frequencies. The general purpose prefabricated breadboard devices may be used up to frequencies of 1 to 5 megahertz depending on circuit complexity. They usually have no particular provisions for applications at very high frequencies. However, there are breadboards available with special provisions for high frequency operation, such as 50 ohm transmission line voltage distribution strips, which are useful to from 50 to 200 megacycles.

Wherever there are critical areas and the breadboarding does not simulate final product operating conditions in these areas, the breadboard fails to provide the final answers. These areas may include heat-sinking, noise pick-up and other environment dependent characteristics. In such cases limited information may be obtained by breadboarding, but final testing must be done in the final and actual environment.

7

How to Devise Complex Systems by Using Simple Basic Elements

Practical ways to convert block diagrams to hardware.

A block diagram provides a simplified overall picture of a functional piece of hardware. A block diagram to be implemented with discrete components leaves many questions as to details unanswered. While the final result may be much the same, as intended by the combinations shown in the block diagram, no two engineers or technicians would fill in the details in exactly the same way.

Integrated circuit block diagrams are much closer to reality. The blocks generally represent standard or at least recognized circuit packages. What may not be complete or accurate is the interfacing, clocking, powering, readout and input/output interface devices. However, these again will often be standard packages once the specific requirement is stated.

Compatibility of components should be examined.

Since no one type of logic is available for all possible logic functions, it is often necessary to mate logic of different types. There are other reasons for combining as well. When logic modules of different logic types are combined there may be a compatibility problem. This arises from the fact that different logic types may use different supply voltages and operate with different input and output logic signal

levels. If the levels and voltages are not the same, an interface device
or circuit may be required.

Noise immunity may spell the difference between success or failure of a system.

Logic signals as applied to integrated circuits consist in 0's and
1's or OFF and ON states. Such signals are applied to the inputs of
gates, inverters and so on, and likewise appear at their outputs. For
positive logic, 0 is some low positive voltage and 1 is a substantially
higher positive voltage. The specifications of the various devices
specify that the 0 will always be below a certain specified voltage and
the 1 will always be above a different specified voltage. This means
that the device inputs will see 0 for any voltage below the specified 0
voltage and 1 for any voltage above the specified 1 voltage. Likewise
the device output provides less than the specified 0 voltage for 0 and
above the specified 1 voltage for 1. These input and output conditions
must hold when more than one device is driven from a given device
output.

The difference between the specified 0 and 1 voltages is called the
indeterminate region since any voltage occurring in this region cannot
be taken to be definitely either an 0 or a 1. The difference between the
specified 0 and 1 which inputs see or accept as 0 or 1 and the actual
outputs representing 0 or 1 of a driving device is called the "noise
margin" or "noise immunity" of the device. This means that the noise
in a given interface circuit will be tolerated if, when it is added to the
actual 0 output or subtracted from the actual 1 output, the instantaneous voltage does not go above the input seen as 0 or below the input
seen as 1 at the input of the driven device.

DTL and TTL logic is generally rated at minimum 1 output 2.4
volts, maximum 0 output 0.4 volts, minimum 1 input 2.0 volts and
maximum 0 input 0.8 volts, so that both high and low rated noise
margins are 0.4 volts. C/MOS logic is generally rated at minimum 1
output 4.99 volts (with 5 volt bias), maximum 0 output 0.01 volt,
minimum 1 input 3.5 volts and maximum 0 input 1.5 volts, so that
both high and low rated noise margins are 1.49 volts. The various
conditions described above are shown graphically for quick comparison in Figure 7-1.

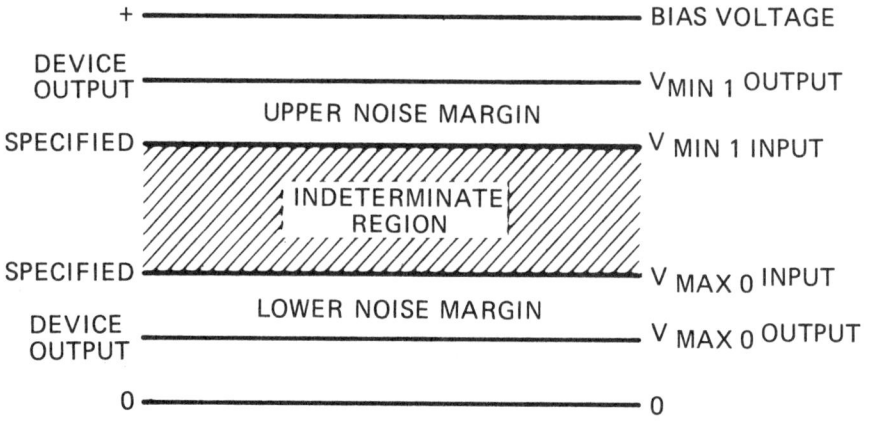

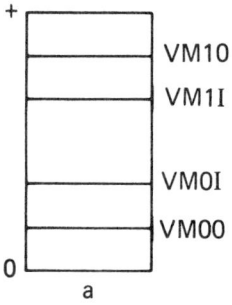

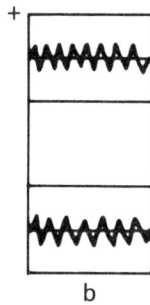

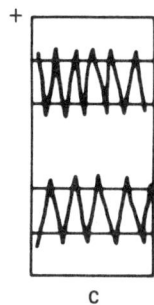

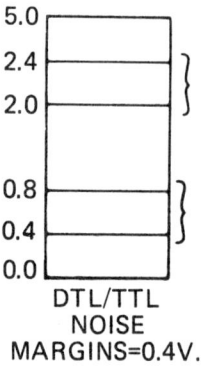

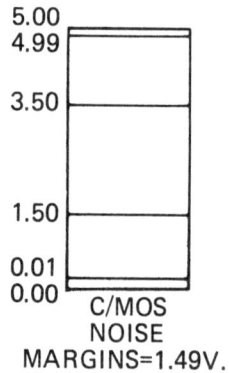

FIGURE 7-1

Understanding interface devices and circuits.

Several important factors must be considered if different types of logic are to be successfully used together. These factors include operating voltages, logic swing, dc input current, output drive capability and noise immunity. Each factor must be considered in turn and provisions made to meet the requirements of both the driving and the driven device. Some devices are more readily interfaced than others and a given order may be easier than the reverse order to accommodate.

The power supply is no problem if both logic types will operate successfully on the same supply voltage. For example, C/MOS devices generally operate from power supplies of 3 to 15 volts, so they can operate on the same power supply as almost any other type of logic. On the other hand, TTL requiring a 5 volt supply interfaced with HTL requiring a 15 volt supply when used together will require two power supplies or a dual voltage supply. The further determinations such as input and output levels are based on the chosen supply voltage.

The next consideration is the logic swing. Bipolar devices driving C/MOS will serve as an illustration. DTL and TTL generally operate on a supply voltage of plus 5 volts which may serve C/MOS as well. The C/MOS device has input thresholds of 1.5 and 3.5 volts, maximum logic 0 and minimum logic 1, respectively. Thus, the DTL and TTL must supply less than 1.5 volts for logic zero and more than 3.5 volts for logic 1 and the resulting noise immunity will be the differences. However, the DTL and TTL logic devices are usually specified to provide 0.4 volts maximum at logic zero, providing a noise margin of 1.1 volts, but 2.4 volts minimum at logic 1. This is then a problem.

DTL or TTL driving C/MOS, for example.

The bipolar logic devices use one of three possible output configurations, namely, resistor pull-up, open circuit or active (transistor) pull-up. With the pull-up resistor (the open collector becomes a resistor pull-up with the addition of an external resistor), a resistor value may be possible to meet all the requirements of fan-out, maximum allowable collector current in the low state (I_L max.), collector-emitter leakage current in the high state (I_{CEX}), power consumption, power-supply voltage and propagation delay time. The fact that the C/MOS requires very little input current usually makes possible the choice of a pull-up resistor to meet all of these requirements. With the active pull-ups rated at 2.4 volts at 400 ua load, the output to a negligible input current C/MOS would be of the order of 3.4 to 3.6 volts and hence no noise margin. Therefore, it is usual to add a pull-up resistor to increase the output voltage of the active pull-up device by shunting the transistor and usual diode junction voltage drops. (See Figure 7-2.)

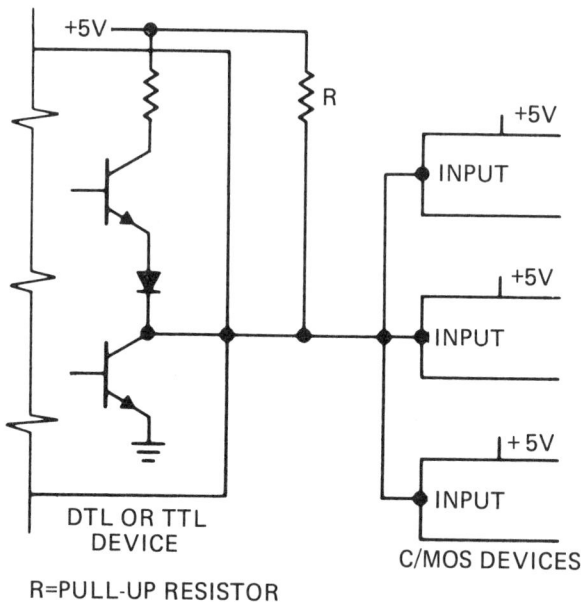

FIGURE 7-2

C/MOS driving DTL or TTL, another example.

The reverse order, i.e., C/MOS driving DTL or TTL logic, presents a different set of problems. These mainly revolve around the ability of the C/MOS to supply and sink the currents required of the voltages required for logic 0 and 1 states. The usual DTL/TTL specifications call for no more than -1.6 ma at 0 state and a maximum of 40 ua at 1 state. Also they specify switching states at voltages from 0.8 to 2.0 volts. Thus, in order to provide a noise margin of 0.4 volts for the

driven bipolar device, the C/MOS must be able to sink the 1.6 ma at 0
logic state voltage of 0.4 volts and -40 ua at logic 1 level of 2.4 volts.
If a single device has insufficient current-handling capacity, two or
more can be paralleled thus mutliplying the capacity of one device by
the number paralleled. For example, four gates rated at 0.8 ma connected in parallel will provide 3.2 ma sink current, enough to drive two
DTL or TTL gates. Other devices which cannot be readily paralleled
must be buffered (added driver stage) between the bipolar and C/MOS
devices. Figures 7-3 may be helpful in visualizing what takes place at
the interface between C/MOS and DTL/TTL logic. Figure 7-4 is a
graphical presentation of input and output voltage levels in saturated
logic (DTL and TTL) and for C/MOS logic.

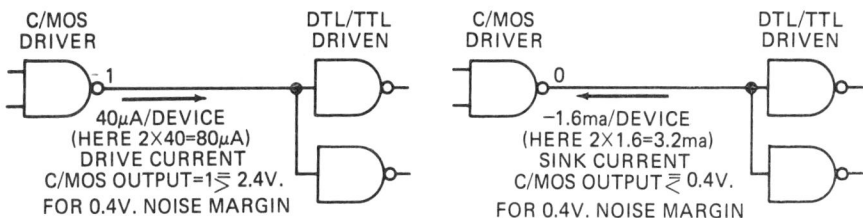

FIGURE 7-3

Interfacing HTL with C/MOS, a further example.

The problem of interfacing HTL (high-threshold-logic) with
C/MOS is much the same as that for the saturated logic discussed
above. While HTL logic is generally operated at supply voltage levels
from 14 to 16 volts, C/MOS can be operated at 15 volts. However, the
HTL devices may present some additional problems. One of these
problems arises from high dissipation in HTL devices and the effects of
temperature on them. Thermal runaway may result if not guarded
against. Compatibility of pulse width and input rise and fall time
should be considered. Also, since the main reason for using HTL logic
is its high noise immunity, any combinational logic should be examined to make sure that noise immunity is not being sacrificed.
Figure 7-4 is a graphical presentation of input and output levels of HTL
and C/MOS logic at 15 volts supply voltage.

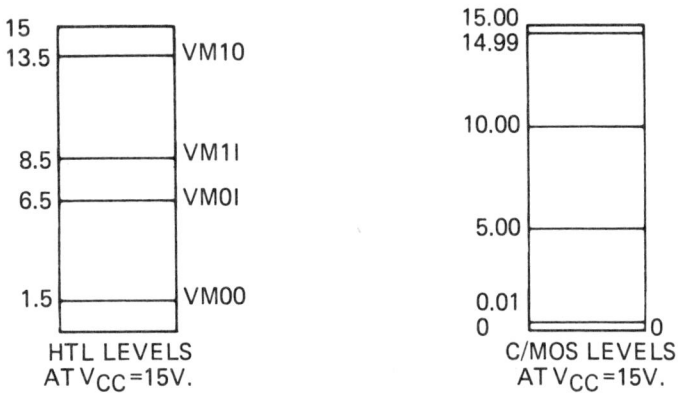

FIGURE 7-4

Interfacing ECL with C/MOS, a still further example.

Interfacing C/MOS with ECL (emitter-coupled-logic) presents
still another problem. This is due to the fact that many ECL devices
operate between -0.9 volts, representing logic 1, and -1.75 volts, representing logic 0. Figure 7-5 shows a simple circuit for interfacing
C/MOS with ECL logic. Other circuits, such as an emitter follower,
can be devised to do the job. Operating in reverse, i.e., driving C/MOS
from an ECL device, is not as easy since it requires a level shifting
circuit. This can be in the form of a linear integrated amplifier or a
level shifting circuit made by the makers of C/MOS logic specifically
for the purpose.

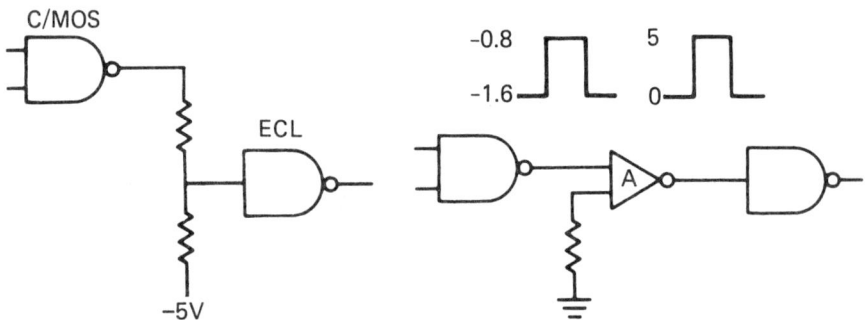

INTERFACING C/MOS AND ECL LOGIC

FIGURE 7-5

Interfacing between other logic types follows the principles set forth above. Translators are often available where required. Between some types, for example DTL and TTL, there is sufficient compatibility so that little or no special provisions need be made. The important thing is to operate each logic type in accordance with its ratings. For most types, these ratings are rather strict, with the exception of C/MOS which will operate over a wide range of supply voltages although sacrificing speed at the lower voltages.

The how and why of input/output devices.

Digital integrated circuit systems, in general, are designed to manipulate information in digital form. When input and output transducers are also digital, input and output devices are simple and straightforward. A digital clock, for example, counts input pulses from an ac power line or crystal oscillator and drives digital output indicators. All of the devices in the system are digital, i.e., on or off at any given instant. The input oscillations and output indicators are also digital and hence can be connected directly to an all-digital processing circuit.

Hybrid systems are widely useful. Analog information is often digitized, manipulated and used in digital form. Conversely, digital information is generated, manipulated and then converted to analog form for utilization. Such systems require conversion; in the first case the conversion is analog to digital (A/D) and in the second case the conversion is digital to analog (D/A). These conversions are actually a mix of analog and digital devices. (See Chapter 5 for details of A/D and D/A converters.)

Meeting fan-in and fan-out requirements.

When combining gates and other logic modules to form systems, it is often necessary to drive more than one device from another device and for one device to receive more than one input. Thus, "fan-in" may be defined as the number of inputs that can be connected to a given logic circuit, and "fan-out" as the number of parallel loads consisting of inputs of the same logic family that can be driven from one output of a given logic circuit. In some cases this is a simple matter presenting no particular problem, while in others it may become quite complex.

Logic families are designed with the idea that various members of one family are to be used together. If this is not to be the case we have the compatibility and interfacing problem discussed elsewhere in this chapter. For the present discussion we will consider the fan-in and fan-out problems logic family by logic family.

There are four main factors to be considered and dealt with when more than one logic circuit is connected to another logic circuit, namely:

1. voltage excursions
2. current requirements
3. capacitance loading
4. noise considerations

A simple basic statement of the fan-out requirement is that sufficient driving current must be available to accommodate all loads in either 1 or 0 condition while maintaining output voltages which do not seriously degrade noise margins. In very high speed logic circuits the further requirement is that added capacitative loading must not seriously degrade the propagation time in the system. In large systems transmission lines may be required for inter-connections to maintain the required speed of response.

First consider fan-out of an RTL gate with a passive pull-up resistor of 640 ohms operated at a nominal Vcc of plus 4 volts. If a 10 percent tolerance is accepted for Vcc, the actual Vcc may be as low as 3.8 volts. If the minimum turn-on voltage is taken as 2.0 volts and a 0.4 volt noise margin is to be maintained, the output of the driving device must supply 2.4 volts into the shunt impedance of the devices to be driven. Now, suppose the input resistors of the RTL are 470 ohms and it is assumed that the Vbe of the driven devices is a maximum of 0.6 volt, then the current required by each input will be (2.4 − 0.6)/470 or 3.8 ma and each device will drop Vcc by 0.0038 × 470 = 1.8 volt. It would seem that this particular circuit can drive only one gate input circuit.

Since RTL is not high speed logic, the added delay of another series device will not seriously affect the system performance and a buffer (driver) can be added where more than one parallel gate input is to be driven. Such a driver may have a collector resistor of 100 ohms, so that extending the above calculations each input will drop the driving voltage only 0.38 volt, and 1.8/0.38=4.74 shows that 4 inputs can be driven in parallel. Another way to make it possible to drive more inputs in parallel, obviously, is to increase the input resistors. If they are made 1000 ohms rather than 470, the above computations, if extended, will show that 10 inputs can be paralleled and driven from a single buffer output while maintaining assumed drive voltage and noise margin.

Driving the RTL inputs to zero presents no problem since the zero output mode of the RTL logic is a saturated transistor and the driven circuits require practically no current sinking. See Figures 3-1 and 3-2 of Chapter 3 for representative circuits.

The situation for DTL differs in that very little if any driving current at logic 1 output is required but some current is required to sink at logic 0 output drive. Thus, DTL is easily made to fan out to a substantial number of inputs at little driving power, either forward (1) or sinking (0). These conditions can be seen by reference to Figure 3-3 where forward drive current is blocked by the input diodes for logic 1 output and the current sink need only handle the base bias current to bring the base to 0 for logic 0 conditions. Thus, fan-out of DTL is relatively simple.

Fan-out of ECL is perhaps the easiest and most natural of all logic families. The inputs are high impedance due to the differential amplifier type of input circuit and thus require little current in either state. Furthermore, the output circuit is active, providing potentially higher drive currents than are possible with the passive circuits of the usual RTL or DTL outputs. (See Figure 3-9.) Typical fan-out of ECL is rated at anywhere from 7 to 90 devices depending on some secondary considerations such as the value of pull-down resistors which may be used.

While the above is a generalization about dc fan-out, the ECL is particularly designed for high speed operation and the shunt capacitance may well become the real systems problem. When several inputs are driven in parallel, their input capacitances add in parallel. This, then, may determine the maximum number of inputs to be paralleled, rather than dc loading considerations. If timing is critical, parallel signal paths may be resorted to. Other special techniques such as the use of transmission line techniques are often indicated.

The driving of several TTL inputs in parallel is rather the inverse of the situation pertaining to RTL. Here the greater current requirement is sink current when outputting logic 0, while at logic 1 output on a small leakage current need be supplied (typically, 1.4 ma per input at logic 0 and 40 ma at logic 1). However, unlike the RTL case, the high current sink requirement is supplied by an active transistor making possible a greater fan-out than with a passive resistor. Typically, TTL is rated to fan out to from 1 to 10 inputs without auxiliary devices. For greater fan-out buffers are available, although care must be exercised in adding series devices in high speed logic circuits, and TTL is considered moderately high speed.

C/MOS logic is probably the most tolerant of a large number of driven inputs since it can supply reasonable output currents (of the order of 1-2 ma) while requiring input currents of the order of 10 pA. At this rate between 100 and 200 inputs can be driven from a single output. Thus, the statement may well be true that C/MOS presents no great problems in the fan-out configuration. As in any case where high speed is important, the multiple input capacitances may well be the final limitation on C/MOS fan-out.

Using fan-in gate expanders.

Fan-in has been defined above as the number of device inputs that can be connected to a given logic circuit. Once fan-out considerations have been taken care of, fan-in becomes more a mechanical problem than anything else. Most logic packages are designed for a predetermined number of inputs per circuit, i.e., two input gates, four input gates, etc. Additional inputs are anticipated in many cases by connections made available for "expansion." Thus, an expandable gate with two inputs has terminals permitting the connection of a gate expander adding, say, four more inputs, making a total of six.

To expand RTL logic, the common collector line and the common emitter line must be made externally available. To these lines is connected an expander comprising a plurality of transistors with common collector and common emitter connections to be connected to the expansion terminals and separate base line connections for the additional inputs. Since the common emitter line is always available as the ground connection only one line need be brought out, namely the common collector line. See Figure 7-6 for a typical circuit of an RTL two-input expandable gate and a three-input expander making possible a five-input device.

The expansion of DTL devices is even simpler than the above. All that is required from the expandable gate is a base circuit connection while the expander comprises a plurality of separate input, common output diodes. See Figure 7-7 for a typical circuit.

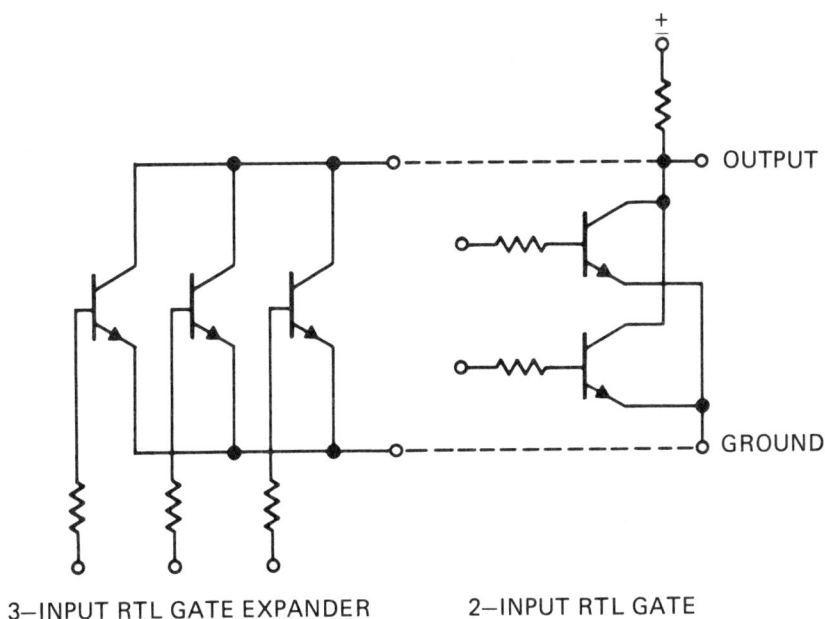

3–INPUT RTL GATE EXPANDER   2–INPUT RTL GATE

FIGURE 7-6

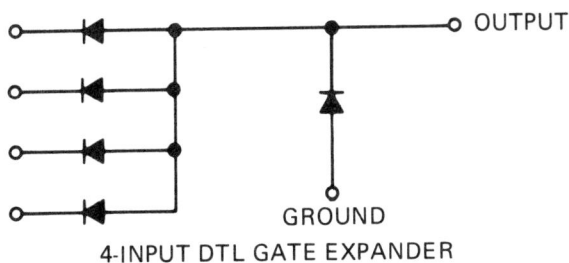

4-INPUT DTL GATE EXPANDER

FIGURE 7-7

Expansion and expanders for ECL follow the same pattern. In this case a plurality of parallel connected transistors like those used in RTL are required. However, in the case of ECL, since ground is not common, both collector and emitter leads must be made available on an expandable device. Figure 3-10 shows an expandable ECL gate and a suitable expander circuit appears in Figure 3-11.

Expansion of TTL devices is just slightly more complicated. Two lines must be brought out from the expandable device which in effect are the common collector and common emitter leads of the transistors driven by the multiple emitter input transistor. Expanders consist in simply one or more emitters and a coupling transistor, as shown in Figure 7-8.

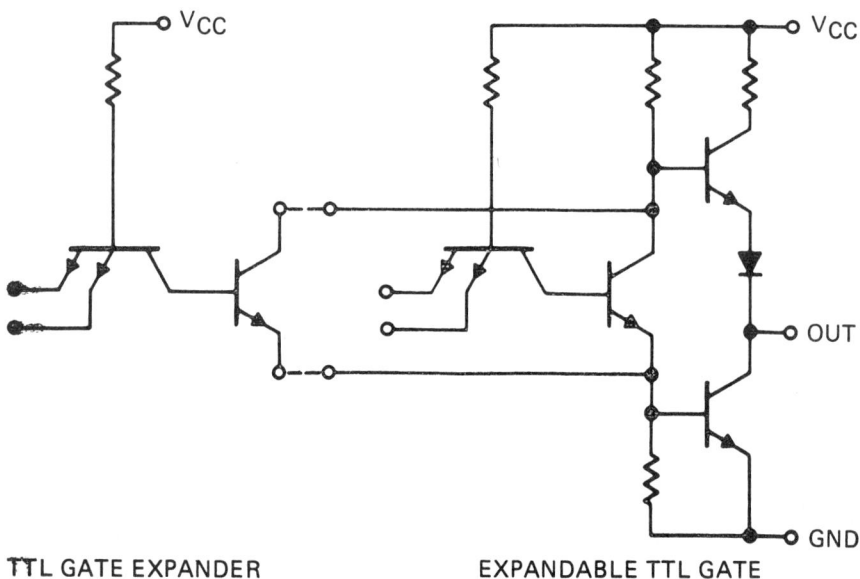

TTL GATE EXPANDER   EXPANDABLE TTL GATE

FIGURE 7-8

Solving problems with opto-coupling.

There are many situations where it is desirable and sometimes imperative that the electrical chain be broken in a system. For example, at the end of a long data transmission line it may be necessary to eliminate accumulated common mode noise. In other cases where common grounds are not possible or are at different potentials isolation may be important. Ground loops may have to be broken. There are many other situations where some kind of isolating interface may be required.

The opto-coupler is the answer in a great many of these situations. The opto-coupler comprises a fast response light source and a light sensitive receiving device (see Figure 7-9). Early opto-couplers used incandescent lamps and photoelectric cells for the light source and receiving device respectively. When the photoelectric cell was followed by a suitable amplifier, the combination could be used to replace relays having typical response times of the order of 1-2 milliseconds. Such devices are still useful in low speed control systems where isolation is required.

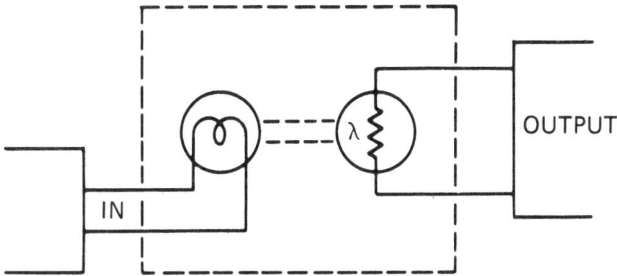

FIGURE 7-9

Isolation is used here to describe a device in which the input circuit is electrically isolated from the output circuit. Any potential or current on the common line is prevented from reaching the output. Generally, differences of potential up to 1-5 kilovolts produce negligible leakage to the output. For example, a typical resistance between input and output devices is greater than 10^{13} ohms. Coupling capacitance between input and output is in the picofarad range, the actual values depending on the package configuration and element sizes.

In order to meet the high speed needs of modern logic circuits, developments have made possible better than a 1000 times improvement over the response time of former relay type isolation circuits. Light-emitting diodes as the light sources and photodiodes or phototransistors as the receiving devices have pushed response speeds to the order of 1-2 megacycles.

Figure 7-10 shows a light-emitting diode (LED) as an input device with a light sensitive transistor as the receiving device, the latter feeding a transistor amplifier stage. Figure 7-11 shows the equivalent circuit where the effect of the light on the phototransistor is to cause a controlled current CC to flow to a virtual base, thereby initiating the transistor action to provide an output.

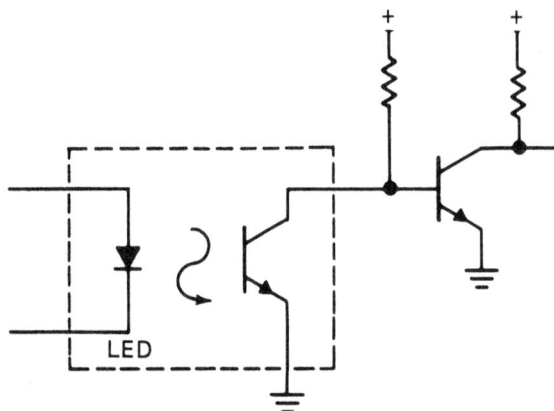

FIGURE 7-10

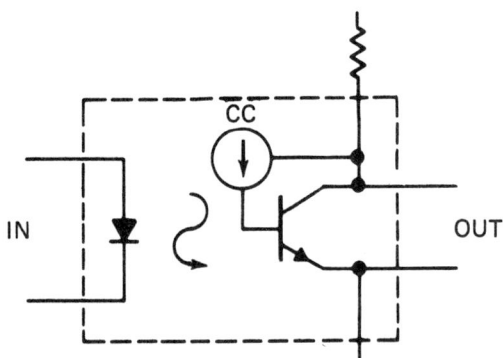

FIGURE 7-11

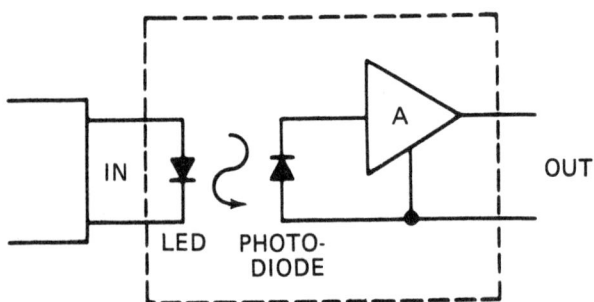

FIGURE 7-12

Figure 7-12 is similar to that referred to above except that here a photodiode is the receiving device and an integrated amplifier is included in the package. This combination can be very sensitive.

As in the case of all interfacing devices, the isolator must be driven properly and its output must drive further circuitry properly. If the terminal device on the input side cannot supply the required LED current, a driver can be interposed as illustrated in Figure 7-13. If the output of the coupler is not compatible with the logic to be driven, a driver like the Schmitt trigger of Figure 7-14 can be used.

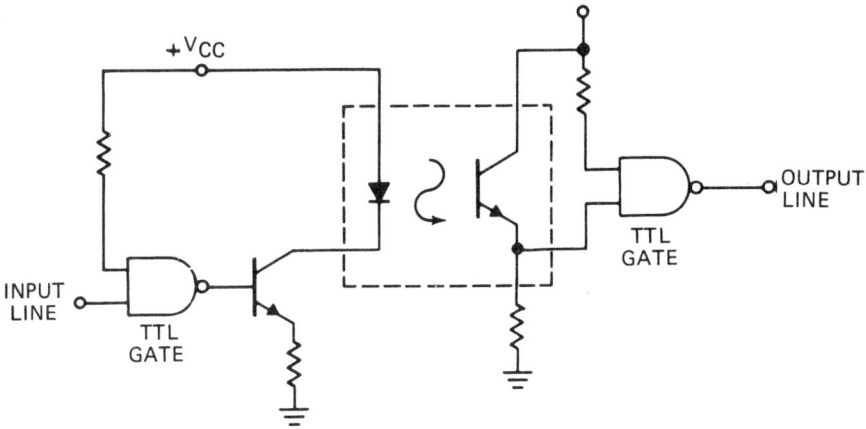

FIGURE 7-13

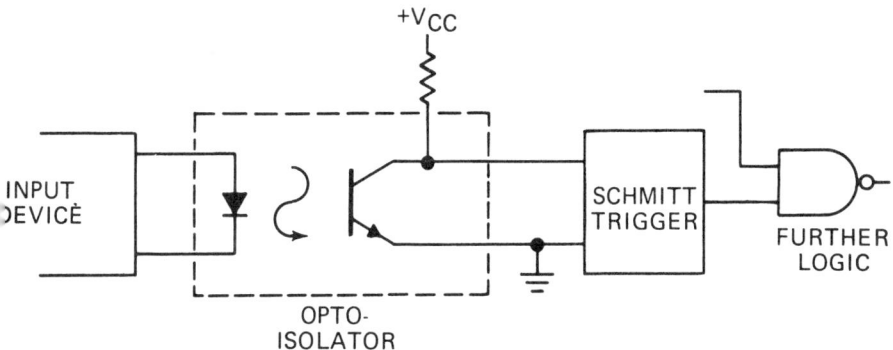

FIGURE 7-14

One of the important inherent features of the opto-coupler found in very few other devices is that there is virtually no feedback from output to input. This solves some of the feedback problems which can require complex conventional solutions. The many uses of these devices have resulted in a substantial development program on the part of some leading manufacturers. We can expect more and better opto-couplers in the future.

Using the important tool of charting to simplify systems.

A chart is a form of truth table which enables one to visualize an equation and often make some simplifications. If a complex function can be simplified, its implementation is likewise simplified.

A chart is simply a group of squares representing the 0 and 1 values of the variables A, B, C, etc., in the equation $X = f_1(ABC) + f_2(ABC)$, etc. Take the case of two variables, A and B. Their chart is shown in Figure 7-15 with the values of the squares indicated. A chart of two variables contains four squares, of three variables eight squares.

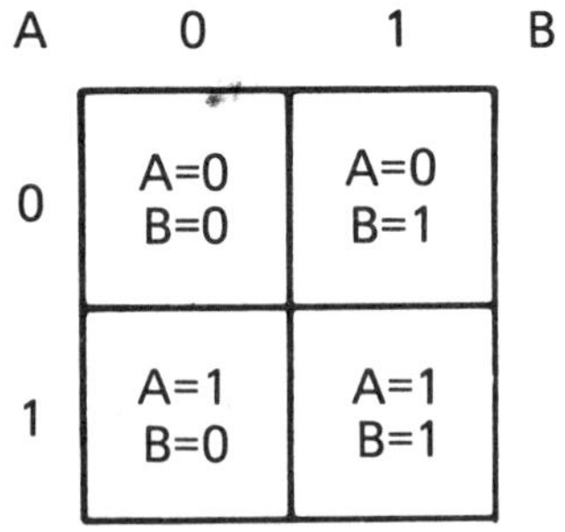

A	0	1	B
0	A=0 B=0	A=0 B=1	
1	A=1 B=0	A=1 B=1	

FIGURE 7-15

Now, to use the chart we select an equation containing the two variables and indicate it on the chart with an open square for 0 ($\bar{A}$ or $\bar{B}$) and shaded for 1 (A or B). Try the equation: $X = \bar{A} \cdot B + \bar{A} \cdot B$. (See Figure 7-16.) We see that two shaded squares are adjacent to each other which indicates that the equation can be simplified. By inspection we can see that the answer is B since either value of A may be used, and the equation simplifies to X = B. We would arrive at the same answer using Boolean algebra but as may be evident from the discussion below, the charting method is often easier to apply in complex situation.

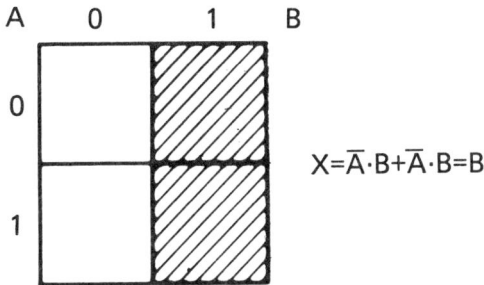

$$X=\overline{A}\cdot B+\overline{A}\cdot B=B$$

FIGURE 7-16

A chart for three variables, A, B, and C, is shown in Figure 7-17 as conventionally drawn. The numbered squares have the following values:

1 = 0,0,0;   2 = 0,0,1;   3 = 0,1,0;   4 = 0,1,1;
5 = 1,0,0;   6 = 1,0,1;   7 = 1,1,0;   and 8 = 1,1,1.

Thus, X = A$\cdot\overline{B}\cdot$C would designate square No. 6; X = $\overline{A}\cdot$B$\cdot\overline{C}$ would be square No. 3; and so on.

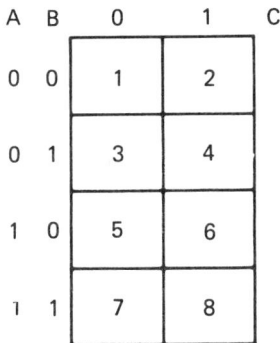

FIGURE 7-17

In Figure 7-18 is shown the representation of the equation $X = \overline{A} \cdot \overline{B} \cdot C + \overline{A} \cdot B \cdot C + \overline{A} \cdot B \cdot \overline{C} + A \cdot \overline{B} \cdot \overline{C}$. To implement this equation directly would require four 3-input AND gates followed by a 4-input OR gate as shown in Figure 7-19. This equation is charted term by term in Figure 7-18. The first term is represented by shaded square No. 2, the second term by shaded square No. 4, and so on. Now, since this is an equation in which the four AND terms are ORed, a chart of the entire equation is simply the sum of all the squares as shown by the fifth chart in the row. It is noted that some of the shaded squares are adjacent to other shaded squares. This means that either value of a variable meets the equation and hence can be dropped, thus leading to the simplification. Adjacent squares 2 and 4 equal $\overline{A} \cdot C$ since either

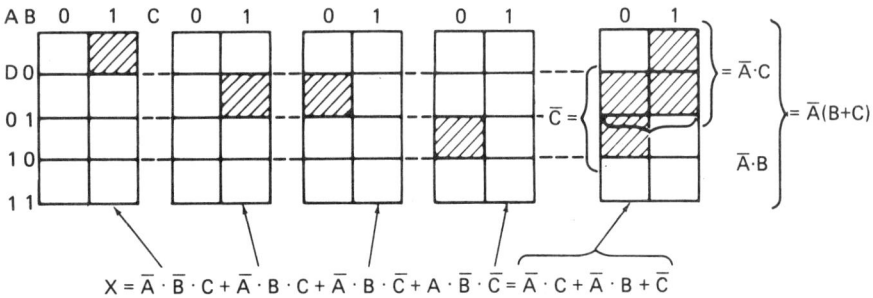

$$X = \overline{A} \cdot \overline{B} \cdot C + \overline{A} \cdot B \cdot C + \overline{A} \cdot B \cdot \overline{C} + A \cdot \overline{B} \cdot \overline{C} = \overline{A} \cdot C + \overline{A} \cdot B + \overline{C}$$

FIGURE 7-18

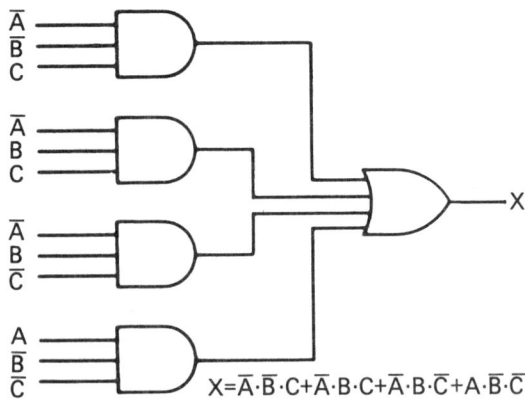

$$X = \overline{A} \cdot \overline{B} \cdot C + \overline{A} \cdot B \cdot C + \overline{A} \cdot B \cdot \overline{C} + A \cdot \overline{B} \cdot \overline{C}$$

FIGURE 7-19

value of B satisfies; squares 3 and 4 equal $\overline{A} \cdot B$ since either value of C satisfies; and squares 3 and 5 equal $\overline{C}$ since either value of both A and B satisfies. The simplified equation becomes $X = \overline{A} \cdot C + \overline{A} \cdot B + \overline{C}$, which can also be written $X = \overline{A} \cdot (B+C) + \overline{C}$. This equation can be implemented by two 2-input AND gates and a 3-input OR gate or by one 2-input AND gate and two 2-input OR gates as shown in Figures 7-20 and 7-21 respectively.

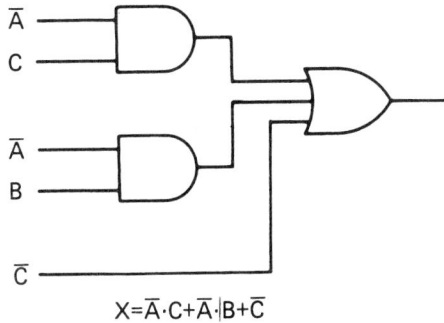

$$X = \overline{A} \cdot C + \overline{A} \cdot B + \overline{C}$$

FIGURE 7-20

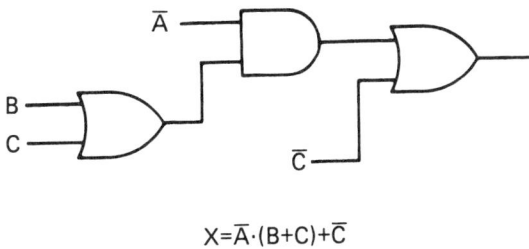

$$X = \overline{A} \cdot (B+C) + \overline{C}$$

FIGURE 7-21

Charting with more than three variables follows the same rules as for two and three as described above. There are some basic rules to go by in any case, as follows:

1. Any two adjacent squares can be looped on a two-variable chart.

   Note: The top row is considered adjacent to the bottom row and the extreme right-hand and left-hand columns are also considered adjacent. Also, when four squares are looped, they can be in one row, one column or 2 × 2, while eight squares must be 2 × 4.

2. Two or four adjacent squares may be looped on a three-variable chart.
3. Any two, four or eight squares may be looped on a four-variable chart.

Not only can equations be simplified and hence their implementation made easier, but it is also possible to go directly from a chart to hardware. Conversion can be made directly to NAND or NOR gates.

Figure 7-22 presents an equation for a series of ANDed and ORed terms together with a chart in which each term is shown as a shaded square. The shaded squares are looped into three loops. One loop represents C, another $\overline{A} \cdot B$, and the third $A \cdot \overline{B}$. These can be combined using NAND gates as shown, providing the resultant $X = C + \overline{A} \cdot B + A \cdot \overline{B}$, which is what was deduced from the chart and which could also be obtained by simplifying the original equation. It will be noted that the technique used is to loop the 1's (shaded squares) and apply to a double inversion series of gates.

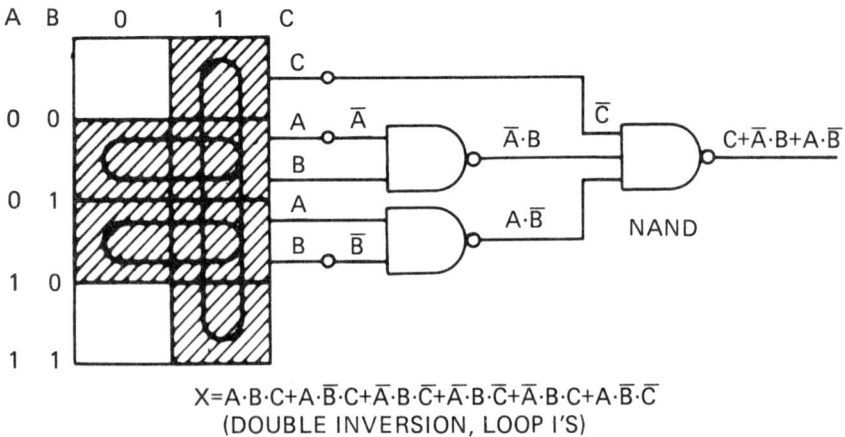

$X = A \cdot B \cdot C + A \cdot \overline{B} \cdot C + \overline{A} \cdot B \cdot \overline{C} + \overline{A} \cdot B \cdot \overline{C} + \overline{A} \cdot B \cdot C + A \cdot \overline{B} \cdot \overline{C}$
(DOUBLE INVERSION, LOOP I'S)

FIGURE 7-22

Figure 7-23 shows another way to reduce a chart directly to hardware. Here, when the equation is plotted, blank squares predominate.

The rule here is to loop the 0's (open squares) and use a single inversion. It is always well to double check the results by simplifying the equation.

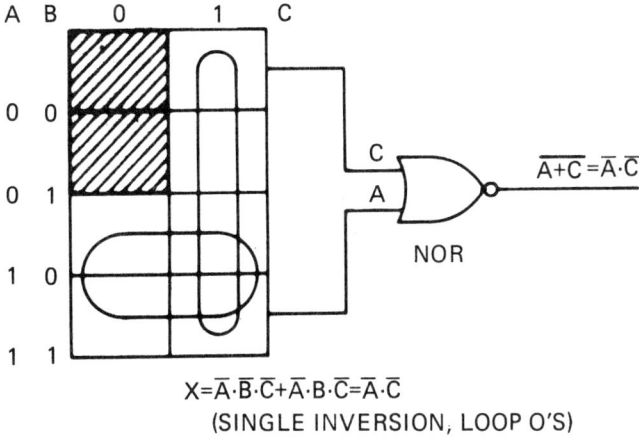

$$X=\overline{A}\cdot\overline{B}\cdot\overline{C}+\overline{A}\cdot B\cdot\overline{C}=\overline{A}\cdot\overline{C}$$
(SINGLE INVERSION, LOOP O'S)

FIGURE 7-23

8

Analyzing

the Fantastic Developments

in Large Scale Integration (LSI)

LSI has created a revolution in the IC field.

The individual diodes and transistors which go to make up an IC have been systematically reduced in size until literally thousands can be integrated on one tiny chip. There are many actual and potential uses for these highly developed ICs. For example, random access memory (RAM) is in demand and would be useful for almost any number of bits if it could be provided reliably and at a viable cost. Another area where LSI has proved its worth is in many devices or systems requiring a variety of functions such as in pocket calculators, electronic watches and automatic cameras. As the individual diodes and transistors are made smaller, factors such as chip purity, masking accuracy, and heat dissipation become more important. As more and more functions are packed on a single chip, cost and yield become increasingly important.

To summarize, by making the active circuits smaller a very large number of them can be integrated on a single chip, making possible a still further reduction in size and cost of complex devices and systems using integrated circuits. As a practical matter the typical LSI employs from 100 up to about 6000 or so transistors and diodes. A device or system requiring a great many more would be provided on two or more LSI chips. One further limitation to the complexity of LSI modules is the number of external connections required in a given case. While as many as 40 terminals are common, it is obvious that a device with several hundred terminals would present some severe handling problems.

Ion implantation is a powerful tool.

LSI is practiced in many ways. No single process provides all advantages in all applications. The CMOS technology has provided some very high density chips (over 6000 transistors and diodes on a 0.1 by 0.1 inch chip), but CMOS circuits are not as fast as TTL logic. Turning to TTL logic, not only is the packing density potential less but the power dissipation is much greater, putting a limit on how many simultaneous functions can be handled on a single chip.

Ion implantation has been used to increase conductivity and hence speed of MOS devices. Ion implantation consists in forming an n-channel of increased conductivity and speed between p drains. A source of high speed boron ions at energies ranging from 80,000 to 300,000 volts is focused on a partially formed and still masked gate structure. The mask and heavy oxide layer of the unfinished structure confines penetration mainly through a thin oxide layer to the channel, thereby increasing its conductivity.

The ion implant MOS/FET has some unique advantages. An ion implanted depletion-mode circuit will carry more current, will respond faster, is compatible with TTL logic gates, requires only one bias source and can be integrated in less area on a chip. Furthermore, by using this method it is possible to integrate depletion-mode and enhancement-mode transistors on the same chip.

What goes into a pocket calculator.

The pocket calculator is an example of LSI put to a use which has found wide acceptance. Figure 8-1 is a block diagram of a simple basic pocket calculator. The heart of the calculator is a single IC which carries at least input circuits (register), an arithmetic circuit for performing the called-for functions (add, subtract, multiply or divide), and a data display decoder (seven-segment output). In order to conserve power, the display is generally multiplexed, calling for a digit multiplexer on the calculator chip. This calculator on a chip is being continually developed, for instance, by adding decimal point circuits permitting a floating decimal or a decimal at a given point, say 2 places. Overflow is generally provided either to indicate that there is an overflow, to stop further entries or to reset the system to zero. Other popular additions have been made to the arithmetic portion so that percentage and/or square root can be directly calculated.

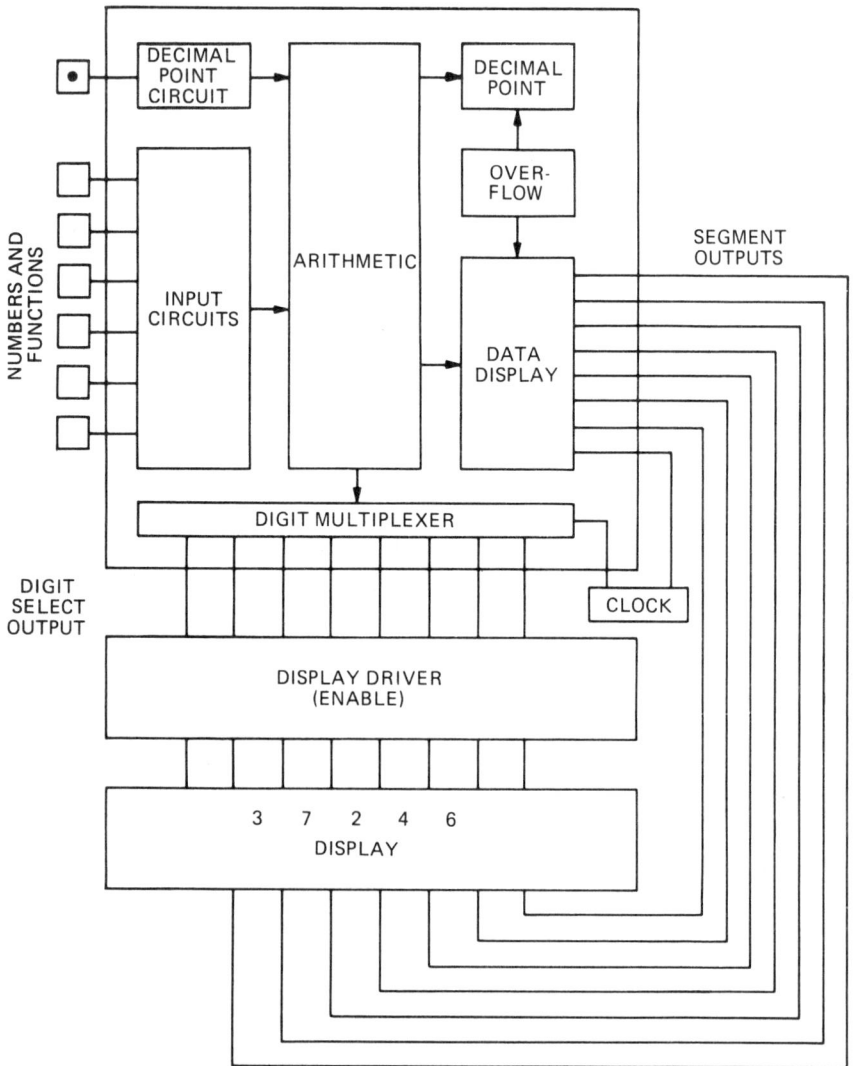

FIGURE 8-1

The function of this calculator chip is to receive and display numbers to be used in the calculations. If the process is to be addition, the addend is entered by sequentially pressing the numbered keys until the complete number is displayed. Entry in the input register may be automatic or it may be completed by pressing the "enter" key. When the enter key is also the "plus" key the addend is entered by pressing "enter +" and then keying in the augend, followed by pressing the "enter +" key again to complete the addition. When there is a separate "equal" key, enter the addend, press plus key, enter augend and display sum by pressing the equals key.

The purpose of the above is to outline the operation of the calculator chip in general and to point out that there are a number of differently organized circuits, each requiring its own particular manipulation.

While the calculator on a chip receives input data and function commands it also puts out display information. The purpose of the display is to show the data contained in the input register and to show the results of the functional commands (add, subtract, multiply or divide). Most displays are seven-segment light-emitting diodes or similar illuminated numerals. The calculator puts out two pieces of information. First, it enables seven output lines going to corresponding segments of the display; and second, it multiplexes each digit in sequence so that only one digit is on at any given instant, thereby conserving power. An internal or external clock pulses the data display and multiplexer providing the sequencing pulses.

The display has been mentioned above. This consists in six or more seven-segment light-emitting numerals plus a decimal point. This display is the main power-consuming part of the calculator. In order to handle this power, the calculator chip may be aided by a display driver which provides the drive for the display.

A most important part of a pocket calculator is its keyboard. The keys on the keyboard are simple contact closures generally connecting a horizontal wire with a vertical wire in matrix form. These keys are operated by finger pressure. A simple contact closed by finger pressure is very likely to bounce, and instead of entering a single clean signal will enter two or more signals. This problem can be met in two ways: (1) the input circuits may contain a delay circuit so that a second signal cannot be entered until a predetermined time has elapsed, or (2) the contacts may be snap-acting with both mechanical and electrical "snap" or toggle. This latter is to be preferred since the signal is clean and the operator can sense by the feel of the key that an entry has been made.

How technology is providing high capability pocket calculators.

Adding a memory chip to the basic calculator permits more complex calculations to be performed. Intermediate or partial answers can thus be stored and brought out for use in latter stages of a computation. Providing more than one memory or register further increases the flexibility of the calculator. Arithmetic and algebraic processing such as squaring, taking roots, and finding sine or tangent may be added in the form of a ROM processor. When the calculator is provided with the capability of using pre-recorded program cards, a vast array of calculations is made possible. The MOS/LSI with the equivalent of 6000 transistors per chip is responsible for extremely wide capabilities in a pocket size calculator. Capabilities which started in large computers and were later incorporated in mini-computers now can be carried in one's pocket. Pre-recorded programs are available in the fields of mathematics, statistics, surveying, medicine and electrical engineering, to name a few. Some examples will illustrate the vast range of subtopics. In mathematics sample subjects include quadratic equations, solution of oblique triangle, unit conversion, Kelvin functions; in statistics they include random number generation, multiple linear regression, parabolic curve fit; in surveying, compass rule adjustment, elevations along a vertical curve, three-wire leveling; in medicine, male vital capacity, Dubois body surface area, blood acid-base status, stroke work; and in electrical engineering, reactance chart, impedance of ladder network, Chebyshev filter design, transformer design, and so on.

Electronic clocks—use ICs.

"Time keeping" is a natural with digital ICs. Many circuits require "clocks" in the form of uniformly spaced pulses. Flip-flops are natural frequency dividers. So, take a stable frequency source, divide it properly and read out the result and you have a clock.

There are two ways in which the input reference frequency can be provided for an all-electronic clock. One is to use the 60- or 50-cycle power line frequency; and the other is to provide a crystal oscillator or similar stable frequency source. Clocks using the power line as their frequency reference will be described since cordless clocks using quartz crystal references are very similar to quartz watches described below.

Figure 8-2 is a block diagram of a conventional all-electronic clock utilizing the power line 50/60 cycles as its source of reference frequency. The block designated "digital clock timer" generally contains all of the essential elements of a clock with the exception of the readout means, driving means for the readout and the power supply (source of dc bias). The timer using the 50- or 60-cycle frequency source contains divider circuits to reduce the 50 or 60 cycles to 1Hz.

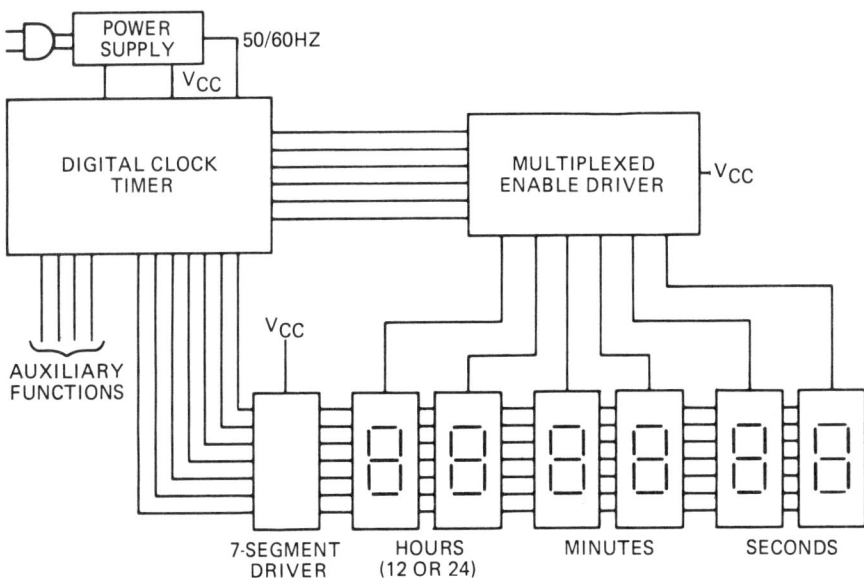

FIGURE 8-2

This one cycle per second is then counted from 1 to 60 and the count is decoded to provide either a BCD or a 7-segment output marking the seconds of time; each time the seconds counter fills, it provides a carry which is counted in a minutes counter from 1 to 60 which in turn is decoded to provide BCD or 7-segment output marking the minutes of time, and each time the minutes counter fills it adds one to an hours counter which in turn is decoded and displayed on the hours display. At the end of 12 or 24 hours, whichever is chosen, the counters are all reset to zero and the process starts all over again.

In order to save components, simplify the wiring and reduce the dissipation in the system, the display is multiplexed. All of the 7-segment display elements are connected in parallel. A multiplexed signal enables a given character in the display and at the same instant enables a driver to that same character causing it to light for an instant. Each segment of each character in the display is similarly enabled in sequence. Thus, only one character is turned on at any given time, thereby saving a substantial amount of power and the resulting dissipation. The multiplex frequency can be chosen high enough to prevent visible flicker but low enough not to cause frequency problems in the circuit. If each segment of the 7-segment characters is to be energized individually, the mux frequency must be at least $20 \times 7 \times 6 = 840$ cycles to avoid objectionable flicker.

Various goodies can easily be added to the basic system; among these are means for slewing the count in order to be able to set the time, a calendar output to show days of the month, alarm clock facilities, and so on. With suitable interfacing, high voltage (large character) displays can be driven. Liquid crystal displays require ac excitation having values between 12 and 50 volts RMS. Figure 8-3 is a block diagram of a typical interface between a BCD source and a 7-segment liquid crystal or other display requiring more than 5 volts to drive. Liquid crystal displays have one big advantage when power is at a premium,

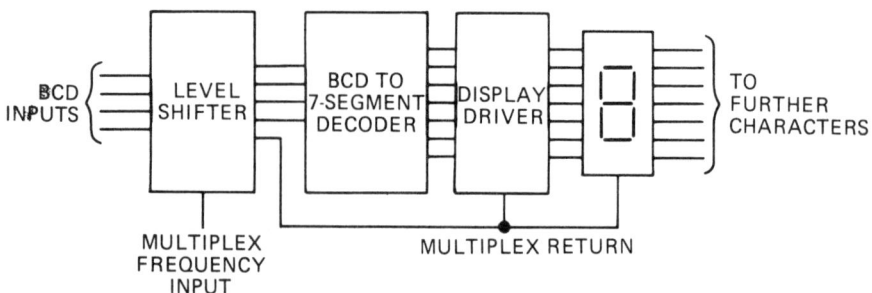

FIGURE 8-3

since a single character has an equivalent circuit consisting of 3 megohms shunted by a capacitance of 50 pF. They have the disadvantage of being visible only due to reflection and refraction of the ambient light.

Here we have a good illustration of LSI and what it is all about. Figure 8-4 is a block diagram of what is inside the digital clock timer LSI module of Figure 8-2. There are fifteen blocks shown, and in an actual set-up there could be several more. Each one of these blocks comprises one or more MSI components. Clocks have been made using the fifteen or so MSI functional modules. This requires a sizeable printed circuit board and, in the case of socketing, fifteen or more IC sockets. However, all of these individual ICs can be combined on a single chip in the form of an LSI single module. The demand being substantial, and the saving in cost and space being considerable, the providing of the circuits in LSI form is a foregone conclusion. The number of external connections is well within the packaging feasibility.

Electronic watches are big users of ICs.

An all-electronic watch or a cordless clock generally uses a quartz crystal as its frequency standard. 32,768 is a common crystal frequency as it is the 15th power of 2 and is a frequency readily supplied by a small quartz crystal.

In order to use the crystal source in a watch (or clock), the frequency is divided by 2 fifteen times, providing a 1 Hz signal. Figure 8-5 is a block diagram of a crystal oscillator watch using a 32,768 Hz crystal. The crystal frequency is divided by 2^{15} providing 1 pulse per second. These pulses are counted from 1 to 60 and a BCD output is provided representing the elapsed seconds; the carry at 1 pulse per minute is counted 1 to 60 and a BCD output is provided representing the elapsed minutes; the carry from this latter counter at 1 pulse per hour is counted 1 to 12 or 1 to 24 and a BCD output is provided representing the elapsed hours. The BCD outputs of the seconds, minutes and hours counters are multiplexed and fed to a BCD to 7-segment decoder, the output of which in turn is connected to all the 7-segment display numerals in parallel. The strobe signal for the multiplexing and for addressing the display numerals is obtained by counting and decoding a 1024 Hz obtained from the fifth stage of the 2^{15} divider to provide six spaced pulses repeated at a rate of 1024/6 or 170 Hz.

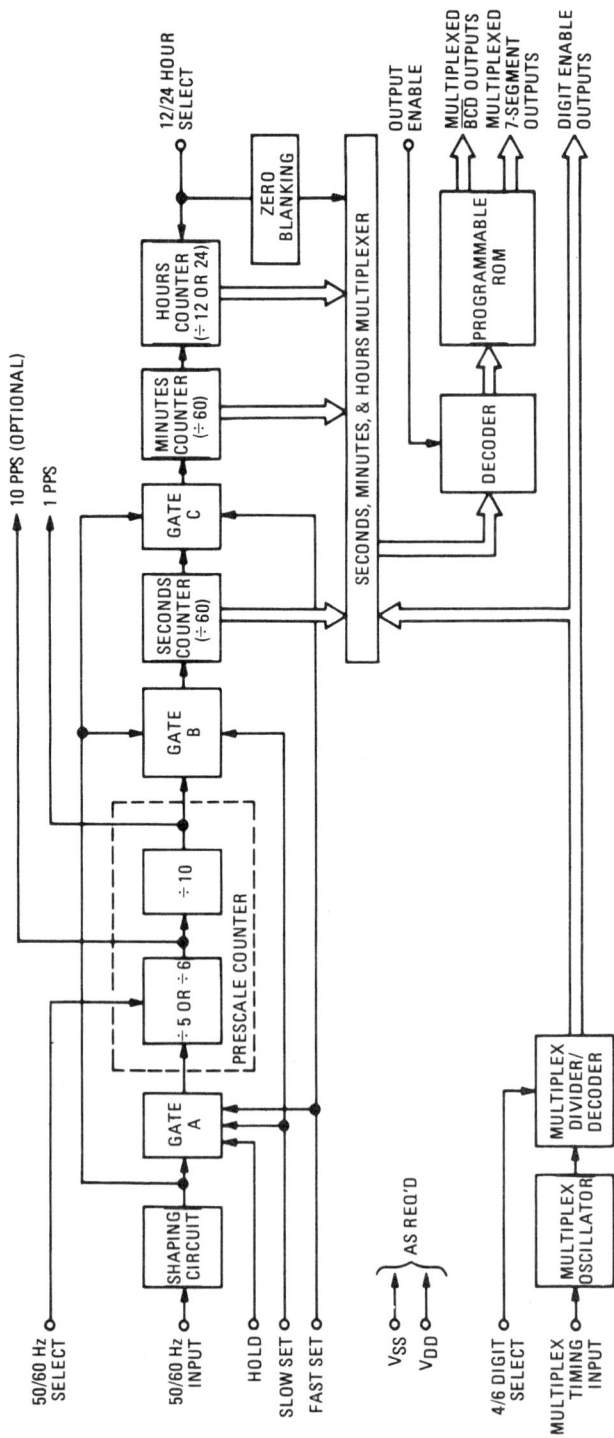

FIGURE 8-4
(National Semiconductor Corp.)

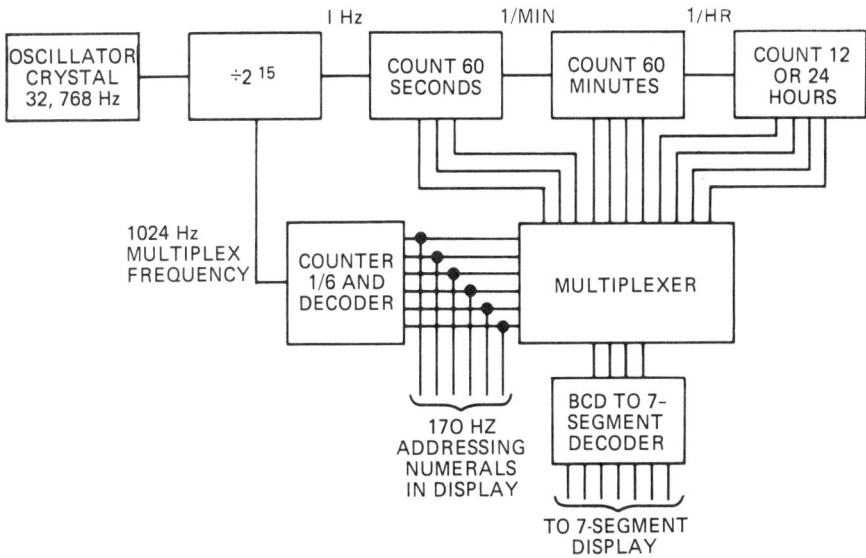

FIGURE 8-5 (National Semiconductor Corp.)

There are other ways of combining dividers, counters, multiplexers and decoders to provide an all-electronic watch or clock, but the combination described above is representative. The all-electronic watch must overcome two major problems; one is size and the other is power. The integrated circuit has been responsible for getting the size to reasonable proportions. The power requirement has favored the liquid crystal display, which is not an ideal answer. Cost has come down considerably since all-electronic watches were first introduced, and no doubt will continue to come down for some time.

LSI RAMs are a challenge to LSI technology.

The ideal memory is one which can store an infinite number of bits and can access any particular bit in zero time. Every memory system which has been devised to date severely compromises one or both of these ideals. However, there are many uses for memories which are far from ideal. The LSI memory has a very limited capacity but is widely used where its small size and fast access capability are needed.

Figure 8-6 is a block diagram of a 16-bit memory organized as a

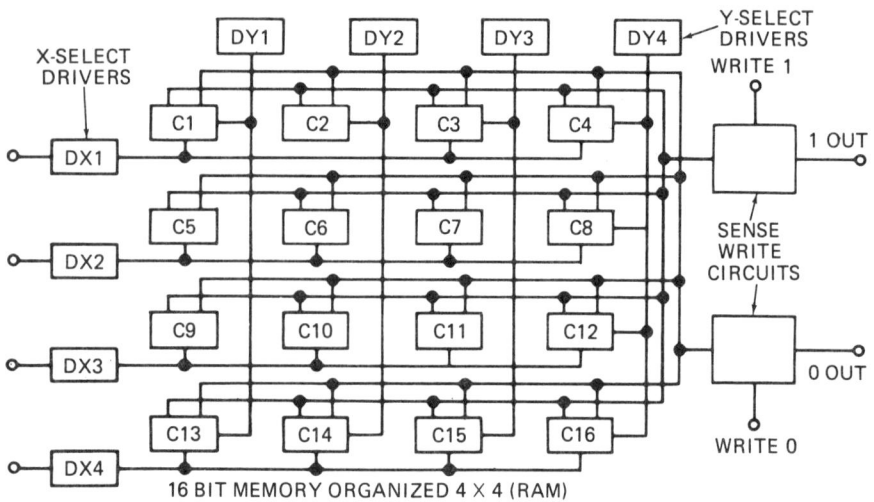

FIGURE 8-6

4-by-4 matrix. The memory cells are bi-stable flip-flops comprising typical TTL configurations. Such a memory, although of very low capacity, can be cycled in as little as 50 ns. Furthermore, with an emitter-coupled sense amplifier, expansion is easily accomplished by wired-OR interconnection with additional memory units. This basic type of memory, like a core-memory, would naturally be a static memory. That is, each memory cell would be individually accessible to read or write, a true RAM with very fast response. The average access time would be the actual access time of any single cell.

By way of comparison, core-memories have typical access times of 750 to 1000 ns, or 15 to 20 times slower than the static TTL memory; the flexible disk has an average access time of the order of 2.5 ms; the standard disk-memory has an average access time of the order of 40 ms; and a 1000-foot section of magnetic tape has an average access time of 25 seconds. The access times are related to the memory capacity—the higher the capacity the longer the access time. The object of technological advance is generally to increase capacity faster than access time or to reduce access time for a given capacity.

The C/MOS integrated circuit, some special forms of TTL, charge coupled logic (CCL), ion implantation have all been used to increase capacity of IC memories while minimizing increases in access time. The small size elements integratable by the above methods result in substantially increased numbers of cells per chip or package. The 256-cell memory was readily produced. This had to be, as are all high capacity IC memories, a dynamic memory, i.e., memory cells are connected in variously organized chains like a counter or register, and are read in or read out by sequencing the information past a read point. This increases access time but keeps total bit members within practical limits (a static memory of 256 bits would require 32 pins just for the X, Y access connections). As the demand increased and the technology improved, significant steps in IC memory capacity were made, e.g., from 256 to 1024 and from 1024 to 4096 bits and so on per chip or package and to a point where practical minicomputers can be made using only IC memory means.

9

How to Specify and Select ICs

Factors determining the selection of ICs.

The first and most important specification for IC gates and flip-flops is the type of logic, i.e., DTL, RTL, TTL, etc. In other words the first decision to be made is to decide on the logic type. The general criteria may be stated in very broad general terms as follows:

1. Ability of a given type to perform the required functions
2. Availability
3. Cost
4. Power requirements
5. Noise immunity
6. Environmental requirements

No attempt will be made to weigh the various factors which enter into the choice of an IC for a given job. Their relative importance can only be judged in the context of the job to be performed. Cost may be an overriding consideration in a mass-produced very comeptitive device while in another situation noise immunity may be the main determining factor.

Each of the above criteria will be considered in some detail below. Good engineering design requires a careful weighing of all factors.

1. Performance requirements include such characteristics as speed of response, interfacing compatibility fan-in and fan-out capabilities, and complex function module availability.

Low speed, non-synchronous logic systems can use practically any type of logic as far as basic performance requirements are concerned. However, as speed requirements are increased the field begins to narrow. RTL, DTL, standard TTL and C/MOS are to be considered slow-to-moderate speed logic devices. Schottky-clamped TTL and ECL are considered high speed logic devices. ECL logic is available in several speed ratings in which speed and power dissipation are traded against each other. The speed requirement is dictated by the device or circuit function subjected to the highest switching rate. In a synchronous system timing, signals and circuit operations must keep in step. However, if this highest speed requirement applies only to a portion of the system, lower speed logic may be used elsewhere. Using a single logic type may simplify the engineering of the system but it may not produce the best overall design.

The rating of logic speed has not been standardized. However, in veiw of available types, I suggest that logic be designated as "high speed" when gate propagation time is less than 10 ns; as "medium speed" for gate time from 10 to 100 ns; and "low speed" for gate times greater than 100 ns.

Where there is a choice, interfacing compatibility may be an important consideration. Logic types which can be readily interfaced permit more flexible system design. Mixing of logic types may help reduce cost and power requirements of a system. In large systems both become increasingly important.

Fan-in and fan-out capabilities are a part of the same picture. Logic types chosen for one reason may inhibit a design due to restrictions on fan-in and fan-out. Another type or a gate expander may be the answer. The transmission gate is another building block which may be useful in this contest.

Not all logic types or families are available for all complex functions. It is not necessary to assume that a given family must be used because of this. It may be quite practical to use the complex function module from one family and interface with another family for the balance of the system.

Good design also dictates that the ICs chosen are optimum for the desired function, i.e., where a two-input AND gate is required, a two-input AND gate is chosen. There are times when an additional input may be provided just in case it may be needed, but generally unused inputs are surplus and may cause problems.

2. Availability is a complex problem. It may mean one thing to the design engineer and quite another thing to the purchasing or production department. If the project is one looking to big production for several years to come, great care must be exercised to use ICs which can reasonably be expected to be available for years to come. This may rule out developmental devices which may not become standard or special devices made by newly formed companies no matter how attractive the performance and price may be.

This may be a good place to point out a common fallacy—that is that all products sold with the same type number are equal. Even the same published specifications are not complete insurance in this regard. There are two ways in which this problem can be met; one is to place a special performance limit on the device, and the other is to so design the system that it will operate satisfactorily with the component variations encountered from production lot to production lot and from manufacturer to manufacturer. The former has been largely responsible for many type numbers which are not unique devices but merely limited forms of standard devices. The latter saves time and money in production and component testing while putting a greater burden on the engineering design work.

For one-of-a-kind or small production, availability becomes no problem where ICs packaged and distributed for experimenters can meet the performance requirements. The cost per IC may be high although mail order houses are very competitive, but this is easily justified in the saving in procurement time. Waiting for parts for experimental work and prototypes is one of the most inefficient uses of engineering time. This is often overlooked by purchasing departments.

3. Cost is an increasingly important consideration as production quantities and competition increase. One real way to reduce cost is to design a system with a minimum number of components. Digital design often lends itself to redesign for simplification. Once a functionally satisfactory sysetm has been worked out, it is often possible to rework it to perform the same functions with fewer components.

NAND and NOR gates are more readily available and generally cheaper than AND and OR gates due to the use of fewer components, and hence design oriented tò these gates can save money.

Saving power saves money, at least in the long run. It also decreases the size of the power supply and may reduce heating significantly. Cooling can be very expensive. Thus, designs using a mixture of types, using low power units where possible, can lead to cost savings.

Sockets can be a source of considerable expense. Their use should be weighed carefully. Expensive equipment requiring maintenance may warrant the use of sockets. Systems made up of plug-in cards with a limited number of components per card probably do not.

4. Power requirements have been mentioned above. Power means heat, and heat may affect performance and reliability or require expensive heat-dissipating devices. Power requirements in digital IC systems have a way of piling up. The 100 milliwatts of one module may not seem like much—but multiplied by 100 it becomes 10 watts and by 1000 it becomes 100 watts. Total dissipation per card and per draw or chassis must be carefully considered in complex systems. Logic type mixes to minimize power dissipation may be a big help in meeting the heat problem.

Regulated power is generally specified for most logic families except for some C/MOS. Regulation increases reliability of performance in two ways. Input and output signals are more predictable and overheating can be more readily guarded against with regulated supply voltages. The reasons C/MOS does not require regulated power supply are that this family is tolerant of wide variations in input and output voltages and there is little danger of overheating due to the very low power consumed. There is one exception, and that is as the signal frequency is increased, heating increases and may become a problem or place a limit on the signal frequency of operation.

5. Noise immunity can be a determining factor in the choice of IC types. Both HTL and C/MOS have relatively high noise immunity voltage characteristics. However, C/MOS exhibits relatively high input impedances so that in some situations the high voltage characteristics may be more than offset. Reducing the impedance with shunt capacitance improves the noise immunity to static type noise but slows the device response substantially.

6. Environmental requirements generally do not favor any particular IC family type since most are made of the same basic material, silicon. Packaging, however, does make a difference. Plastic packages are least resistant to environmental conditions. Hermetically sealed ceramic and metal packages are the most resistant. If heating is a problem, C/MOS or low power forms of other types may be indicated. ICs meeting MIL specifications (military) are subjected to a series of conditioning and performance tests to insure higher than average reliability. Such ICs may be indicated where environmental problems are encountered.

Some Specifications Are More Important Than Others

1. The package content and IC family designation. (Four 2-input NOR gates: RTL, DTL, TTL, ECL, HTL or CMOS. Is the device interchangeable in a given class with other ICs in the same class? If not compatible with a specific class, other specifications must define the device.)
2. Power supply (and bias, if any) voltage minimum and maximum normal and limits.
3. Typical fan-outs.
4. Typical propagation time. (Give range if dependent on some factor such as supply voltage.)
5. Package power dissipation outputs 0's.
6. Package power dissipation outputs 1's.
7. Any characteristic which is non-standard or which cannot be deduced from the above items 1-6.
8. Case type. (This will generally be given with reference to a particular manufacturer's designation if not an industry standard.)
9. General class such as MIL, hobby or industrial.
10. Price. (Due to the changing price pattern in this field, current prices are difficult to provide.)
11. Date. (The date of manufacture is very often useful to know but also often omitted.)

Digital ICs are not all alike, making selection important.

There is a tendency to regard all digital ICs bearing the same type number as being complete equivalents. This is due to the fact that digital ICs are essentially switches and to the assumption that all switches are alike. In fact, digital ICs made by the same manufacturer and even from the same batch may vary considerably from device to device. In simple non-critical applications there will generally be no problem. However, in critical applications there may be serious problems. One answer is to specify more characteristics. Another is to apply closer tolerances to standard specifications. The types of characteristics which are subject to variation and hence may cause problems include propagation time, package dissipation with frequency, package dissipation vs. duty cycle, noise immunity vs. input or output impedance, and so on. Applications which were not contemplated by the module design may impose requirements not anticipated and hence not specified or controlled. This is the kind of situation which results in a special form or specially tested standard form of IC. If the application is sufficiently important a new device may be born.

Application notes can help provide applications instructions.

Comprehensive application notes are a valuable addition to the bare specifications of an IC device. This is especially true of what might be called special devices, i.e., devices mainly provided to fill a particular function. However, the special device designed and described for a special purpose very often suggests uses not contemplated by the original designer or the application notes-writer. Here one may be forced to extrapolate from the intended use and applications descriptions to predict whether or not the device will perform satisfactorily in the new application. Figure 9-1 is a reproduced page from a typical Application Note, this one being from RCA.

Special testing is sometimes called for.

Testing of any component can well account for a substantial portion of the selling price. How the testing is divided between manufacturer, supplier and user is an economic decision. When standard parts do not meet the need, special testing may be required, or perhaps a better answer is redesign to permit the use of standard parts. Not only are special parts and special testing items of increased cost, but the need for special parts also makes field servicing a greater problem.

ℝℂᴀ

**Solid State
Division**

Digital Integrated Circuits

**Application Note
ICAN-6101**

The RCA COS/MOS Phase-Locked-Loop
A Versatile Building Block for Micro-Power
Digital and Analog Applications

by David K. Morgan and Goetz Steudel

INTRODUCTION

Phase-locked-loops (PLL's), especially in monolithic form, are finding significantly increased usage in signal-processing and digital systems. FM demodulation, FSK demodulation, tone decoding, frequency multiplication, signal conditioning, clock synchronization, and frequency synthesis are some of the many applications of a PLL. The PLL described in this Note is the COS/MOS CD4046A, which consumes only 600 microwatts of power at 10 kHz, a reduction in power consumption of 160 times when compared to the 100 milliwatts required by similar monolithic bipolar PLL's. This power reduction has particular significance for portable battery-operated equipment. This Note discusses the basic fundamentals of phase-locked-loops, and presents a detailed technical description of the COS/MOS PLL as well as some of its applications.

REVIEW OF PLL FUNDAMENTALS

The basic phase-locked-loop system is shown in Fig. 1; it consists of three parts: phase comparator, low-pass filter, and voltage-controlled oscillator (VCO); all are connected to form a closed-loop frequency-feedback system.

With no signal input applied to the PLL system, the error voltage at the output of the phase comparator is zero. The voltage, $Vd(t)$, from the low-pass filter is also zero, which causes the VCO to operate at a set frequency, fo, called the center frequency. When an input signal is applied to the PLL, the phase comparator compares the phase and frequency of the signal input with the VCO frequency and generates an error voltage proportional to the phase and frequency

difference of the input signal and the VCO. The error voltage, $Ve(t)$, is filtered and applied to the control input of the VCO; $Vd(t)$ varies in a direction that reduces the frequency difference between the VCO and signal-input frequency. When the input frequency is sufficiently close to the VCO frequency, the closed-loop nature of the PLL forces the VCO to *lock* in frequency with the signal input; i.e., when the PLL is in lock, the VCO frequency is identical to the signal input except for a finite phase difference. The range of frequencies over which the PLL can maintain this locked condition is defined as the *lock range* of the system. The lock range is always larger than the band of frequencies over which the PLL can acquire a locked condition with the signal input. This latter band of frequencies is defined as the *capture range* of the PLL system.

TECHNICAL DESCRIPTION OF COS/MOS PLL

Fig. 2 shows a block diagram of the COS/MOS CD4046A, which has been implemented on a single

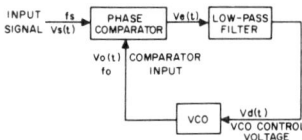

Fig. 1— Block diagram of PLL.

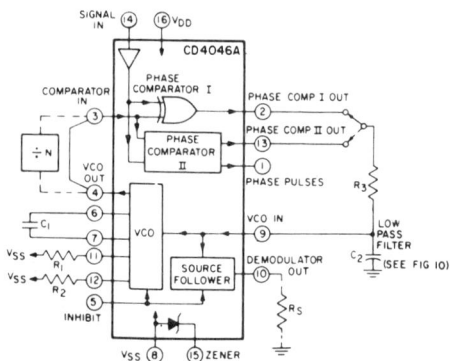

Fig. 2— COS/MOS PLL block diagram.

FIGURE 9-1

ICAN-6101 ───────────────────────────────

monolithic integrated circuit. The PLL structure consists of a low-power, linear, voltage-controlled oscillator (VCO), and two different phase comparators having a common signal-input amplifier and a common comparator input. A 5.4-volt zener is provided for supply regulation if necessary. The VCO can be connected either directly or through frequency dividers to the comparator input of the phase comparators. The low-pass filter is implemented through external parts because of the radical configuration changes from application to application and because some of the components are non-integrable. The CD4046A is supplied in a 16-lead, dual-in-line, ceramic package (CD4046AD); a 16-lead, dual-in-line, plastic package (CD4046AE); or a 16-lead flat-pack (CD4046AK).

Phase Comparators

Most PLL systems utilize a balanced mixer composed of well-controlled analog amplifiers for the phase-comparator section. Analog amplifiers with well-controlled gain characteristics cannot easily be realized using COS/MOS technology. Hence, the COS/MOS design shown in Fig. 3 employs digital-type phase comparators. Both phase comparators are driven by a common-input amplifier configuration composed of a bias stage and four inverting-amplifier stages. The phase-comparator signal input (terminal 14) can be direct-coupled provided the signal swing is within COS/MOS logic levels [logic $0 \leqslant 30\%$ (VDD-VSS), logic $1 \geqslant 70\%$ (VDD-VSS)]. For smaller input signal swings, the signal must be capacitively coupled to the self-biasing amplifier at the signal input to insure an over-driven digital signal into the phase comparators.

Phase-comparator I is an exclusive-OR network; it operates analogously to an over-driven balanced mixer. When using this phase comparator, the signal- and comparator-input frequencies must have 50-percent duty cycle. With no signal or noise on the signal input, this phase comparator has

an average output voltage equal to VDD/2. The low-pass filter connected to the output of phase-comparator I supplies the averaged voltage to the VCO input, and causes the VCO to oscillate at the center frequency (fo). The frequency range over which the PLL remains locked to the input frequency (lock range) is close to the theoretical limit of ±fc. The range of frequencies over which the PLL can acquire lock (capture range) is dependent on the low-pass-filter characteristics, and can be made as large as the lock range. Phase-comparator I enables a PLL system to remain in lock in spite of high amounts of noise in the input signal.

One characteristic of this type of phase comparator is that it may lock onto input frequencies that are close to harmonics of the VCO center-frequency. A second characteristic is that the phase angle between the signal and the comparator input varies between 0° and 180°, and is 90° at the center frequency. Fig. 4 shows the typical, triangular, phase-to-output, response characteristic of phase-comparator I. Typical waveforms for a COS/MOS phase-locked-loop employing phase-comparator I in locked condition of f_0 is shown in Fig. 5.

Phase-comparator II is an edge-controlled digital memory network. It consists of four flip-flop stages, control gating, and a three-state output circuit comprising p and n drivers having a common output node as shown in Fig. 3. This type of phase comparator acts only on the positive edges of the signal- and comparator-input signals. The duty cycles of the signal and comparator inputs are not important since positive transitions control the PLL system utilizing this type of comparator. If the signal-input frequency is higher than the comparator-input frequency, the p-type output driver is maintained ON continuously. If the signal-input frequency is lower than the comparator-input frequency, the n-type output driver is maintained ON continuously. If the signal- and comparator-input frequencies are the same, but the signal input lags the comparator input in phase, the n-type output driver is maintained ON for a time equal to the phase difference. If the signal- and comparator-input frequencies are the same, but the signal input leads the comparator input in phase, the p-type output driver is maintained ON for a time equal to the phase difference. Subsequently, the capacitor voltage of the low-pass filter connected to this type of phase comparator is adjusted until the signal and comparator input are equal in both phase and frequency. At this stable operating point, both p- and n-type output drivers remain OFF, and the signal at the phase-pulses output is a 1,

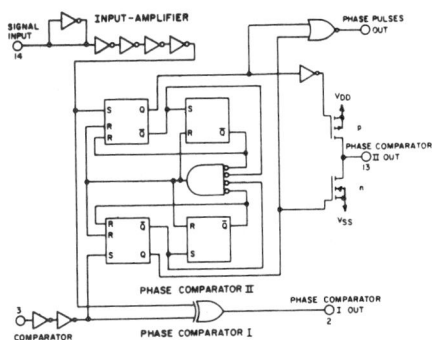

Fig. 3— Schematic of COS/MOS PLL phase-comparator section.

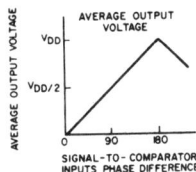

Fig. 4— Phase-comparator I characteristics.

10

Simple and Sophisticated Ways of Testing Digital Integrated Circuits

How digital IC testing differs from analog circuit testing.

Testing digital integrated circuits differs from testing analog circuits in many ways. The very important linearity criteria of analog circuits have no meaning in digital technology. Digital circuits have two states; i.e., they are "on" or they are "off." They don't even have to go all the way on or all the way off to operate perfectly in the circuits for which they are intended. Speed of operation, power handling ability, dissipation and noise immunity are the most important characteristics of digital ICs. In low speed circuits, of these, only the power and noise characteristics are important. Fan-in and fan-out are important in some applications.

While the basic characteristics of an IC are matters of basic design, testing is important to determine whether or not a particular unit is up to its specifications. Slight differences from unit to unit are normal although IC units of a given batch are surprisingly uniform. On the other hand, acceptance limits are quite liberal, so units in which some characteristics differ considerably cannot be considered unacceptable or unlikely to perform as intended.

Some characteristics may change substantially with temperature but without exceeding acceptable limits. Many tests need only determine a maximum or a minimum where the corresponding minimum or

156

maximum has no significance. For example, the output of a gate representing logic 1 may be rated as 2.4 volts minimum. As long as the logic 1 output is 2.4 volts or greater, it is considered a good unit. Actually, it could be 3 to 4 volts wtihout being abnormal in any way. Many specifications can be so loose that even different types of logic can meet them.

Thus, the importance of testing digital ICs is not to see how good they are but to see if they are within specifications or not. Erratic behavior, even when the excursions of response do not exceed the specification limits, is more than sufficient reason for rejection of an IC module. Lack of uniformity between components such as gates in a multiple gate package is suspect. It may indicate contamination due to a crack in the encapsulation. Such a package is quite likely to deteriorate further with time. Non-uniformity or a completely inoperative gate may indicate a local defect in the chip. This may not affect other parts of the package. However, it is a defective module.

The table in Figure 10-1 shows a comparison between the specification limits for a 7400 (TTL) NAND gate and the actual measured values of input and output voltages. These values have been drawn graphically in Figure 10-2 for a better visualization of the way such gates operate. The actual noise margins are also illustrated. The noise margins are the differences between the signal that turns a gate on and the one that turns it off, measured first from the on signal and then from the off signal. It can be seen from this figure how the more the actual signals stay within the specified limits the greater the actual noise margins will be. So, in the case of noise margins, there is a practical advantage with circuits which exceed the specifications.

SPECIFIED VS. MEASURED VALUES OF 7400 (TTL) NAND GATE		
QUANTITY	SPECIFIED	MEASURED
LOGICAL 1 INPUT REQUIRED	2.0V. MIN.	1.4 VOLTS
LOGICAL 0 INPUT REQUIRED	0.8V. MAX.	1.0 VOLT
LOGICAL 0 OUTPUT VOLTAGE (-400μA LOAD)	2.4V. MIN.	4.2 VOLTS
LOGICAL 0 OUTPUT VOLTAGE (16 MA SINK)	0.4V. MAX	0.25 VOLT
SHORT CIRCUIT CURRENT	-18 MA MIN. 55 MA. MAX.	29 MA.

FIGURE 10-1

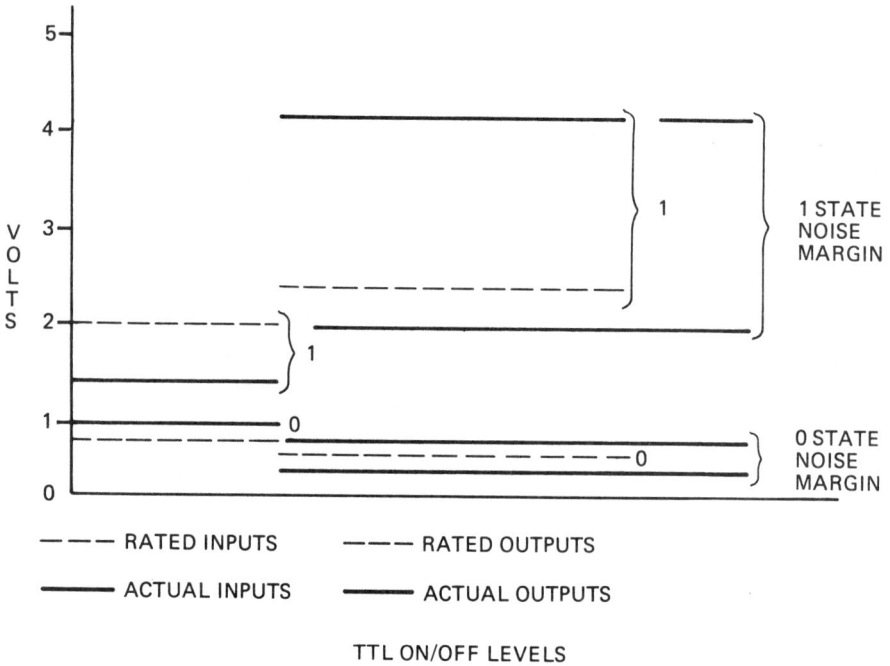

TTL ON/OFF LEVELS

FIGURE 10-2

In testing components for high speed systems, the various time parameters must be measured. Not only is this important in order to determine the suitability in terms of response speed, but it is also important that delay times are a known factor. High speed systems are generally synchronous sytems and require matched time delays in components in order that pulses be in step at various points in the system.

Figure 10-3 is a graphic presentation of a hypothetical input and resulting output pulse as well as a second output pulse illustrating transient response. The significant time intervals associated with the pulse are labeled. The rise time, defined as the time required for the pulse (input or output) to increase an amplitude from 10 percent to 90 percent of its final value, is labeled t_R. The corresponding fall time (90 percent down to 10 percent) is labeled t_F. The propagation delay time, the time between the 50 percent level of the input pulse and the resulting 50 percent level of the output pulse, is labeled t_P. The pulse duration at 90 percent of maximum amplitude or greater is labeled t_D.

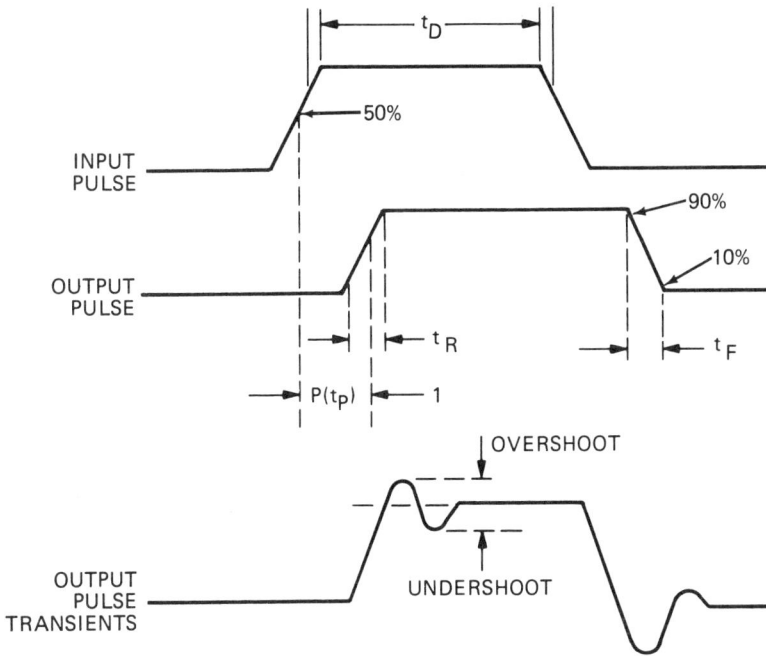

FIGURE 10-3

In high frequency systems, responses may be affected by even small values of inductance resulting in a transient response as shown in the lower diagram of Figure 10-3. The characteristics of such a response find their important components in the overshoot and undershoot as shown. The overshoot is the percentage by which the leading edge of the pulse response exceeds the normal full pulse level. The undershoot is the following half-wave measured by the percentage by which its amplitude falls below the normal full pulse level.

The time parameters of the pulse response can best be observed and measured on a dual beam cathode ray oscillograph. Figure 10-4 shows the basic set-up for pulse response measurements or observation. The pulse generator must be capable of providing a pulse fast enough and with satisfactory shape to exercise the device to be tested. The rise time of the pulse from the pulse generator must be substantially shorter than the rise time of the device to be measured or such

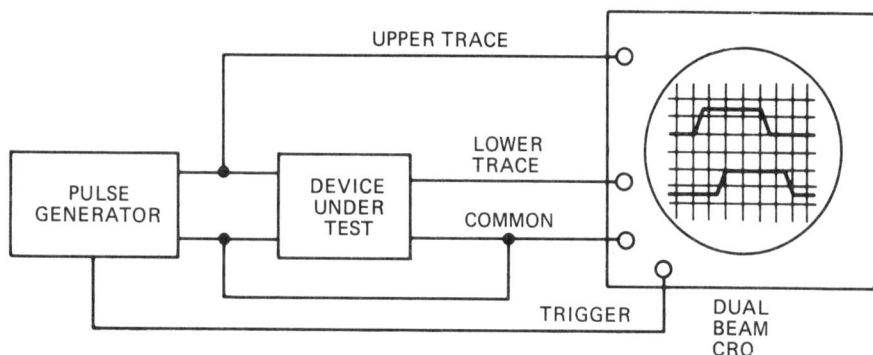

FIGURE 10-4

response will not be measurable since the output rise time will be that of the pulse generator output.

Complex units such as registers, memories, full adders, data selectors, decoders, analog to digital converters, digital to analog converters, and so on, require more complex testing devices and procedures. Useful testing devices are capable of making a number of unique checks, particularly to show the progression of signals through a multistage device.

Figure 10-5 is a diagramatic representation of the way in which one digital analyzer operates. The top line shows the clock pulses being used for timing purposes. The next lower line shows a row of synchronized data pulses at one-half the clock rate (labeled "DATA A"). The second line from the bottom, labeled "LED A," represents a row of light-emitting diodes connected to light whenever the leading edge of a clock pulse finds a high (1) data A signal. The darkened circles represent LED's on. Thus, the 1st, 3rd, 5th, and so on LED's are lighted, showing high data A pulses at corresponding clock pulse leading edges. Now, assume that the device being tested is a frequency divider, count by two or other device yielding an output at one-half the pulse rate of its input. The third line on the diagram labeled DATA B represents the sensed output of such a device, while the last line of the diagram represents the LED B lamps which will light, showing the clock leading edges at which the output (B) is high (1). This represents the expected pattern and that provided by a perfect device. Improper response results in some other, or no, LED pattern for the output (B).

Testers for incoming inspection, final testing or fault location are

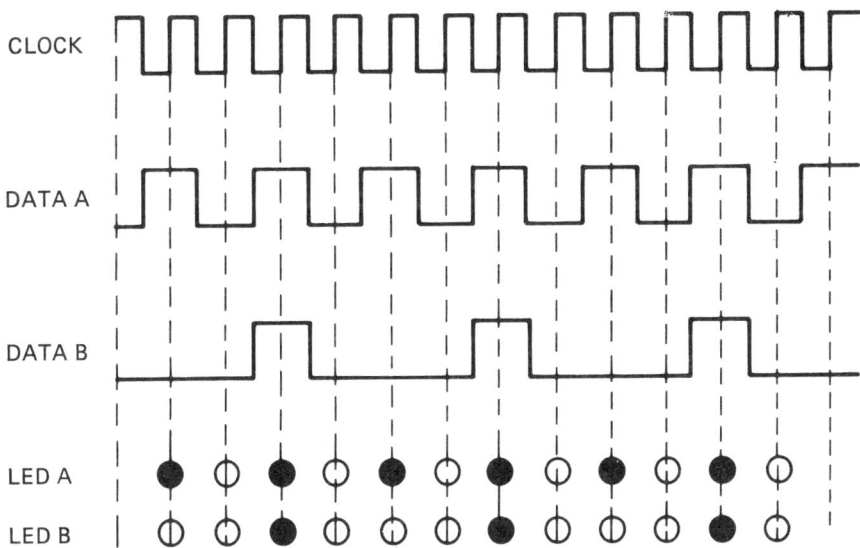

FIGURE 10-5

generally provided with a test clip which fits over the IC under test and makes contact with each of its pins. Testers of this type are relatively expensive but may be necessary if a substantial volume of testing is to be done.

Where the amount of testing is relatively small or infrequent, a single probe type tester may· be sufficient for the purpose. Such probes can be quite versatile in indicating digital circuit states. Indicating d.c. states will show whether a pin of a module (input or output point) is at logic 0 or logic 1. Pulses, however, on a visual readout device must last long enough for the eye to detect or they must be stretched. The pulse-detecting probes employ a one-shot multivibrator having an "on" time of, say, 200 milliseconds, a time sufficient for visual detection.

Figure 10-6 is a block diagram of a probe type tester having three LED readout indicators: one for logic zero, one for logic one, and one for the presence of pulses. The amplifier at the input is provided to prevent loading circuits under test. Power should be provided for the probe indicators for the same reason. This circuit may be the basis for a homemade probe or it may be compared with available commercial probes.

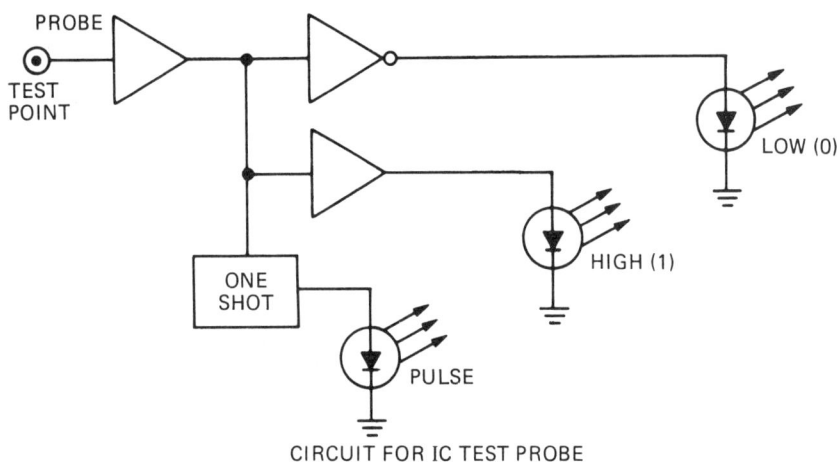

CIRCUIT FOR IC TEST PROBE

FIGURE 10-6

Setting up, operating and testing individual or simple combinations of ICs is helpful in understanding them and is a good basis for testing individual units. A useful set-up for a wide range of tests comprises a box carrying two IC DIP sockets on its cover and a box with input and output means. Each box is provided with a complete set of pin-tip jacks permitting setting up any combination of input and output circuits quickly.

Figure 10-7 shows the cover layout and Figure 10-7A the internal circuitry of the input/output signal box. Figure 10-8 shows the cover layout of the IC socket box (Figure 10-8A shows the connections). Power is to be supplied by means of batteries or a small regulated power supply of appropriate voltage for the type of logic to be tested. When experimenting or testing, a regulated power supply with current limiting has the advantage of preventing destructive currents from flowing due to a wrong connection. Better still, a power supply with a current meter will more quickly show wrong connections or defective ICs.

The jacks and push-button switches of the test box of Figure 10-7 and 7A are labeled for a J-K flip-flop although gates and other types of flip-flops can be tested as well. Now, for example, to test a J-K flip-flop, leads with pin tips on both ends are used to connect the J, C, K, and S and R terminals of a J-K flip-flop to be tested and plugged

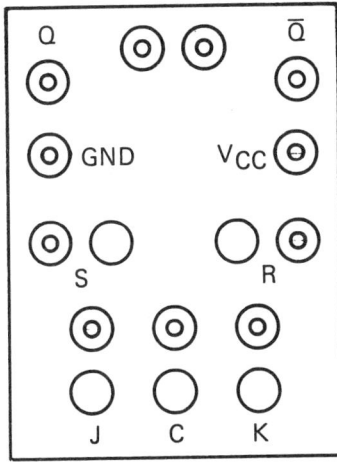

IC TEST BOX

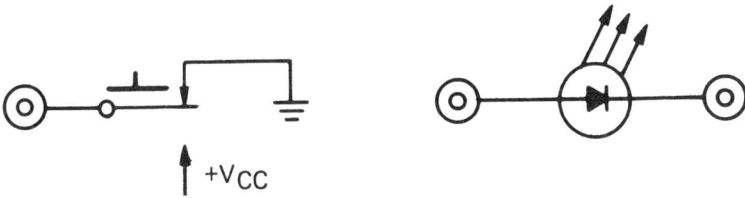

INTERNAL CIRCUITS

FIGURE 10-7

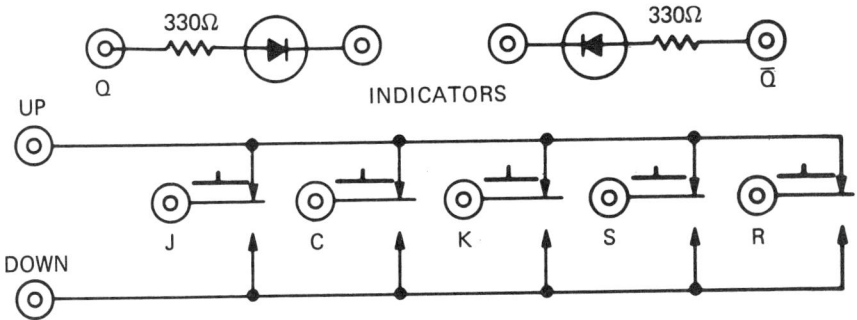

INDICATORS

FIGURE 10-7A

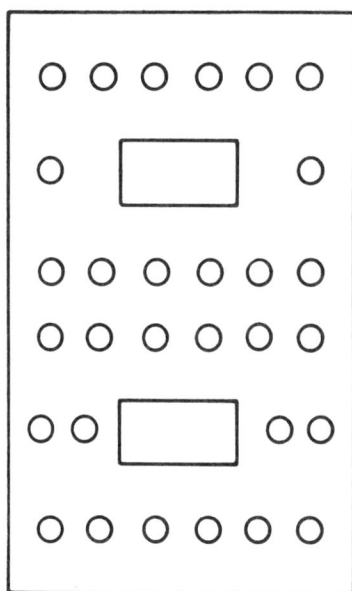

IC SOCKET BOX

FIGURE 10-8

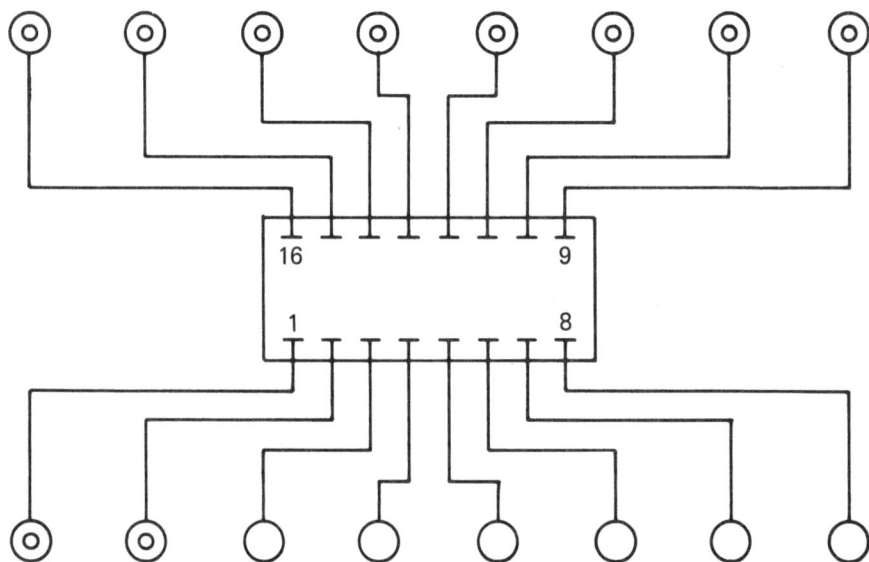

FIGURE 10-8A

into the socket of the box shown in Figures 10-8 and 8A. The Vcc and return from the bias voltage source are connected to the designated terminals (for example 4 and 11 respectively for a type 7473) of the IC socket box. Now, if a pulse-5 volt source is connected to the "UP" jack and ground return to the "DOWN" jack, the J, C, K, S and R jack connections will all be at plus-5 volts if no push-buttons are pressed. In order to ground any one or more of these inputs, the corresponding push-button is pressed. The resulting outputs can be observed by connecting the Q and $\bar{Q}$ terminals of the IC socket to Q and $\bar{Q}$ on the jack box and connecting both LED return terminals to ground.

Many other ICs can be tested in a similar manner. Using the two-socket box combinations of ICs can be experimented with or tested. For example, a 7-segment LED numeral can be placed in one socket, a BCD to 7-segment decoder in the other socket, and the push-buttons can be connected to provide binary count inputs. The combination can be used to test or experiment with a wide range of digital ICs.

How to test to specifications.

IC specifications, particularly those for digital circuits, may seem very loose to one familiar with electronic component specs in general. The primary reason for the seemingly loose specifications is that digital circuits are essentially switches which are programmed to be either on or off. The on and off states are critical in themselves and become significant mainly because digital devices must work in harmony. For example, a gate may provide an output of not more than 0.4 volt to represent logic 0 and not less than 3 volts to represent logic 1. Generally, this gate will be connected to another gate. If the second gate switches from 0 to 1 and vice versa at 1.5 volts, the first gate will have a margin (noise margin) of $1.5 - 0.4$ or 1.1 volts for 0 and $3.0 - 1.5$ or 1.5 volts for 1. These margins are important since they insure proper gate switching and, in addition, a safety factor called the noise margin.

Gate inputs, except for those employing C/MOS devices, draw current or supply current to the driving gate or other device. Hence, it is important to know if the compatible input/output conditions will be maintained in an actual circuit where currents as well as voltages are involved. The current considerations are particularly important in systems employing many gates fanned in or out of other gates.

The current drawn from the bias source is important since it determines the total dissipation for the device. Individual devices may require little power but IC systems typically employ large numbers of devices, thus multiplying the dissipation to a point where the system dissipation may become a serious problem. Short-circuit current is also important since it is an indication of the ability of the device to withstand a short-circuit.

Before the limits can be finally set it may be necessary to determine the effects of time and temperature. Noise margins, total dissipation and other characteristics may deteriorate at high temperatures or drift with time.

All of the above factors may be called the d.c. characteristics of the system and its components. They are relatively simple to determine using a voltmeter and ammeter.

Complex devices may require sequence and pulse-testing devices as described above. Synchronous systems may, in addition, require propagation time analysis using pulse generators and oscilloscopes.

With this introduction, a typical specification sheet for a simple gate would be as follows:

FUNCTION	MIN	TYPICAL	MAX	UNIT
Input to switch output to 0	2	—	—	volts
Input to switch output to 1	—	—	0.8	volts
Logical 1 output voltage	2.4	3.3	—	volts
Logical 0 output voltage	—	0.22	0.4	volts
Input current at output 0 (per input)	—	—	−1.6	ma.
Input current at output 1 (per input)	—	—	40	μa.
Supply current at output 0 (each gate)	—	3	—	ma.
Supply current at output 1 (each gate)	—	1	—	ma.
Short-circuit current	−20	—	−50	ma.
Propagation delay output 0 to 1	—	18	29	ns.
Propagation delay output 1 to 0	—	8	15	ns.

In some cases the minimums are the most significant values while in other cases it is the maximums. Typical values only are not definitive and should not be accepted as a specification. The above specification sheet may be satisfactory as a general indication of the performance of a given device and may serve as the reference when "testing to specifications." However, it does not supply all the information which may be necessary for evaluation purposes. It is assumed that the single spec sheet above is to be taken as that pertaining to 25° C. operation. For a more complete evaluation the specifications should provide limits at rated maximum and minimum temperatures, say, at +85° C. and −30° C. Type testing may be carried out to check these values but incoming inspection tests, for example, would not be necessary at other than the nominal +25° C. temperature.

The above discussion has assumed TTL logic. It applies to other logic types by merely changing to the appropriate voltage levels and polarities.

The six test circuits for making the d.c. tests are shown in Figure 10-9 through Figure 10-14. The first of these, Figure 10-9, shows the test set-up for measuring the input voltage required on an input to insure the output going to logical 0. The circuit is also designed to measure the logical 0 output voltage when sinking a predetermined current—typically 16 ma for TTL logic.

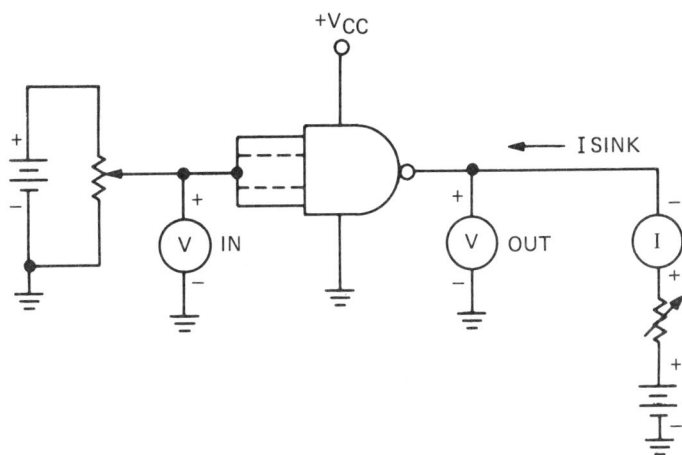

LOGICAL 0 OUTPUT VOLTAGE

FIGURE 10-9

The second of these, Figure 10-10, is for making the corresponding measurements for the opposite condition, i.e., logical 1 output. In this case the output will generally drive a load so that the direction of the current is reversed. The variable resistor is used to provide a load drawing a specified current—typically, 0.4 ma for TTL logic.

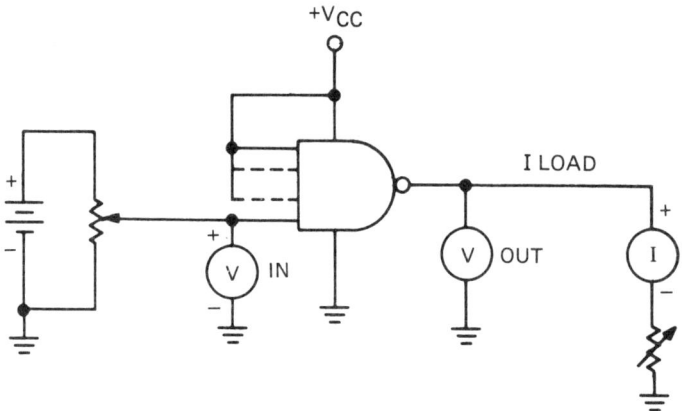

LOGICAL 1 OUTPUT VOLTAGE

FIGURE 10-10

The next, Figure 10-11, is the set-up for checking input current for logical 0 input. It is a sink in TTL logic since current flows from the input terminal to ground. Each input terminal should be tested separately. While the absolute value of current for this test may vary considerably from manufacturer to manufactuer and still be acceptable, the currents for a given unit should be very nearly the same for each input. An input supplying a significantly different current from the rest in the unit can well be taken to indicate a defective unit.

The following circuit, Figure 10-12, is a similar circuit for input current at logical 1 input. The input current meter is reversed since at logical 1 the inputs draw current (TTL logic). Each input again should be tested separately and they should show uniformity in a given unit.

To continue, Figure 10-13 shows the set-up for measuring short-circuit current. A current limited power supply is indicated for this test

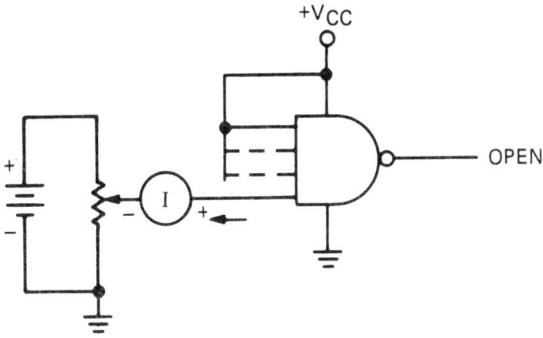

LOGICAL 0 INPUT CURRENT
(TEST EACH INPUT SEPARATELY)

FIGURE 10-11

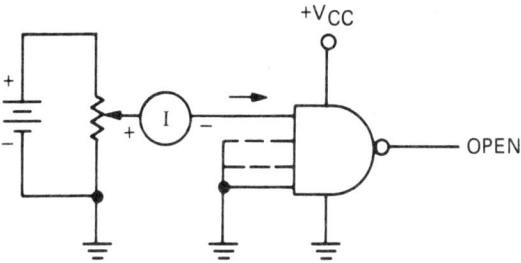

LOGICAL 1 INPUT CURRENT
(TEST EACH INPUT SEPARATELY)

FIGURE 10-12

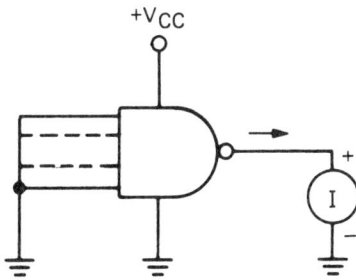

OUTPUT SHORT-CIRCUIT CURRENT

FIGURE 10-13

since without such limiting a short from Vcc to the output could burn out the output current meter.

The final set-up, shown in Figure 10-14, is for measuring the current drawn from Vcc supply. This should be done for logical 1 output and for logical 0 output, all gates to be in same condition. If the current is specified for each gate, the measured current will be N times the rated current per gate where N is the number of gates per package.

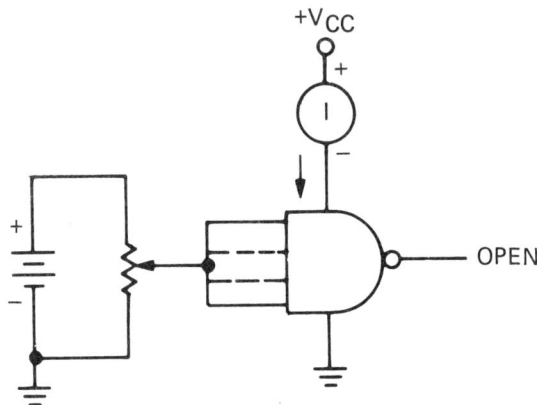

SUPPLY CURRENT ALL GATES
LOGICAL 1 OR LOGICAL 0

FIGURE 10-14

Simple IC testers for simple IC devices.

Users of only a few ICs can hardly afford sophisticated IC testers which typically run into thousands of dollars each. For the small user, simple test set-ups can be made to carry out at least most common tests. Tests have been described above covering most characteristics on a unit-by-unit basis. A simple socket box like one of those shown above will serve to provide input and output access to the IC under test. A little imagination can show how to combine tests by switching. Switching levels can be set or monitored by means of Schmitt trigger circuits and LED indicators. Many tests can be simply on a go-no-go basis. See the description of Figures 10-7 and 10-8 above.

Figure 10-15 shows the lay-out of a different type of test box. Here two DIP sockets are provided, one for a reference IC and the other for an IC which is to be compared. There are two ideas here. One is that two ICs with the same specifications can be compared to find how they may differ, or to locate a defective gate, for example. The other is that an unknown IC may be identified by comparison with a known IC. This box is provided with a series of lever switches, one SPDT switch with center off for each pin for both ICs. When the lever is thrown to the A socket, the resistance of a particular terminal is measured against all of the remaining terminals by means of an ohm-

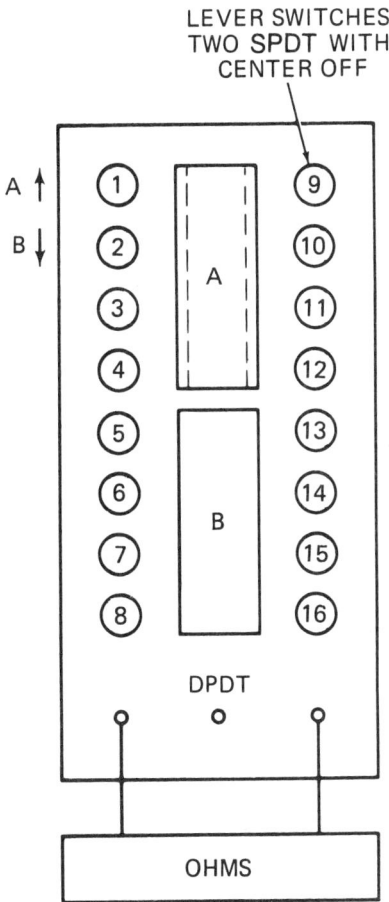

FIGURE 10-15

meter. When thrown to the B socket the same measurement is made for the IC being compared. The ohmmeter is connected through a double-pole double-throw switch so that the test voltage impressed on the IC circuit can be reversed.

A portion of the switching circuits is shown in Figure 10-16. This test box is useful in small quantity testing as it quickly shows whether or not a given IC is substantially identical with a reference IC. It also locates a defective IC such as a shorted or open gate circuit, out-of-specs values or variations among units from the same batch. It can also be used in trouble-shooting where a defective IC is suspected.

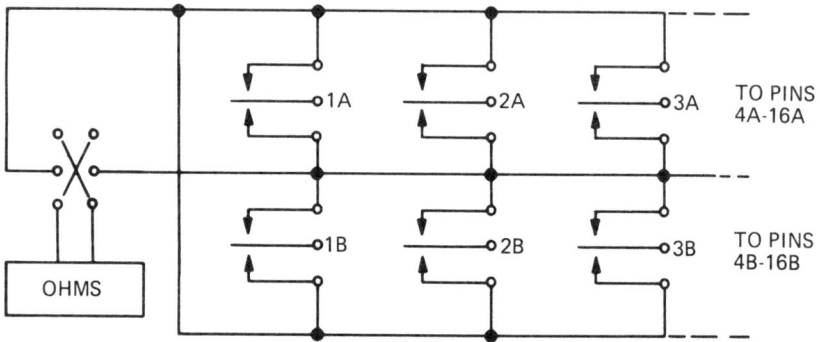

FIGURE 10-16

Logic probes give more information.

Simple in-circuit testing of digital ICs can be done with a logic probe. This is a simple hand-held probe having one or more indicator lights near its probe tip as shown in Figure 10-17. An external connection is made to the circuit to be tested, generally Vcc and GROUND. The probe tip is placed in contact with the IC pins one at a time to determine their signal condition. One type of probe employs a single light. A bright glow indicates "1" or "high"; no glow indicates "0" or "low"; a dim light indicates an open; and a flashing light indicates a pulse train. The flashing light is provided by a stretch-out circuit for slowing pulses to a point of good visibility.

Other probes use somewhat modified indicators as, for example,

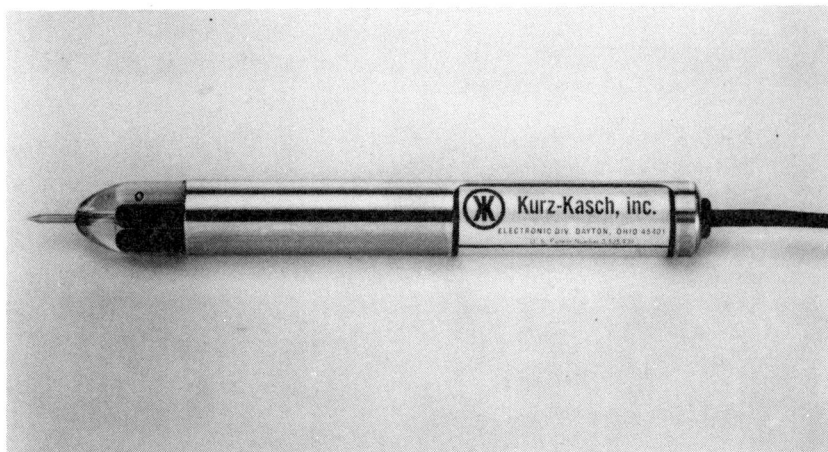

FIGURE 10-17

different colored lights to indicate the different conditions. All of these logic probes are practical only in analyzing simple circuit conditions of one or a few ICs.

Logic clips examine several circuits simultaneously.

The states of all 14 or 16 pins of a DIP IC can be observed simultaneously with a logic clip. This device is equivalent to 14 or 16 probes all connected and functioning simultaneously. The clip by means of suitably spaced teeth or individual clips is connected to all pins of the IC to be tested. A light is provided for each pin so that the state of all pins can be observed at the same time.

A more sophisticated clip has provisions for making a comparison between a reference IC and the IC being tested. A known-to-be-good reference IC is plugged into the head of the clip; the test prods are connected to the pins of the in-circuit IC being tested; the circuit is exercised; and the clip indicators show which pins if any of the IC under test are performing differently from those in the reference IC. The bias for the reference IC is derived automatically from the IC being tested by means of a simple diode selector. A sophisticated clip is shown in Figure 10-18. A block diagram of a complete comparison type clip is shown in Figures 10-19 and 10-19B.

FIGURE 10-18 (Fluke Tendar)

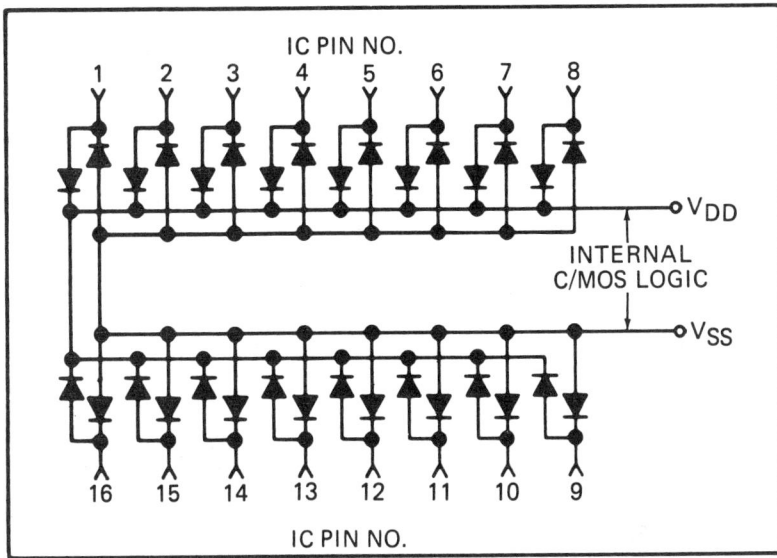

FIGURE 10-19 (Fluke Tendar)

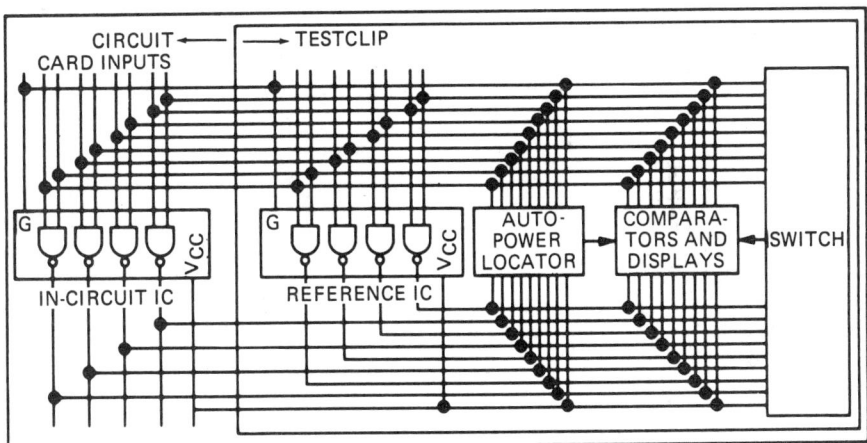

(ELECTRONICS NOV 8, 1973)

FIGURE 10-19B (Fluke Tendar)

When and how to use logic analyzers.

Logic analyzers are directed to the examination of streams of serial data. With such a device it is possible to capture, store and examine the logic sequences from calculators and other ROM-controlled systems. Serial data transmissions, the output from disc storage and many other types of serial data may be studied so that malfunctions can be detected and cured.

The Hewlett-Packard Model 5000A Logic Analyzer will be used to show how a logic analyzer operates and its several modes of operation. The front panel of the Model 5000A is shown in Figure 10-20.

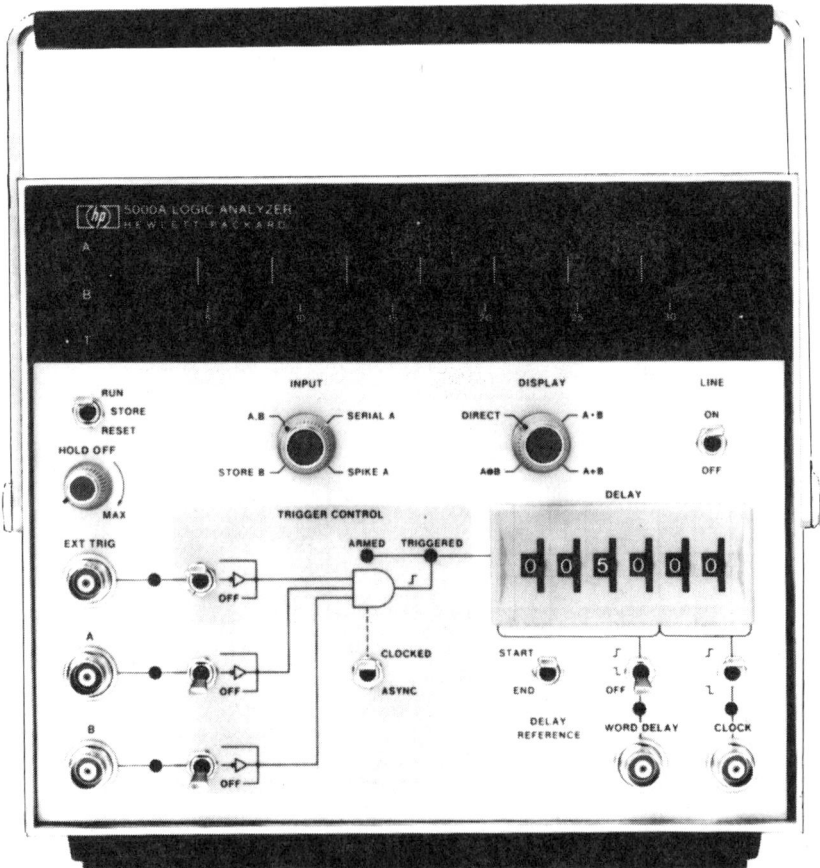

FIGURE 10-20 (Hewlett-Packard Journal)

Indication is provided by means of two rows of 32 light-emitting diodes for indicating up to 32 sequential states of two inputs (A and B).

The analyzer operates under clock control, data being sampled at either edge of the clock pulses, and detected high states are indicated by a lighted LED. The clock of the system being studied is applied to the clock input of the analyzer. Figure 10-21 shows the light pattern resulting from the illustrated data applied to input A.

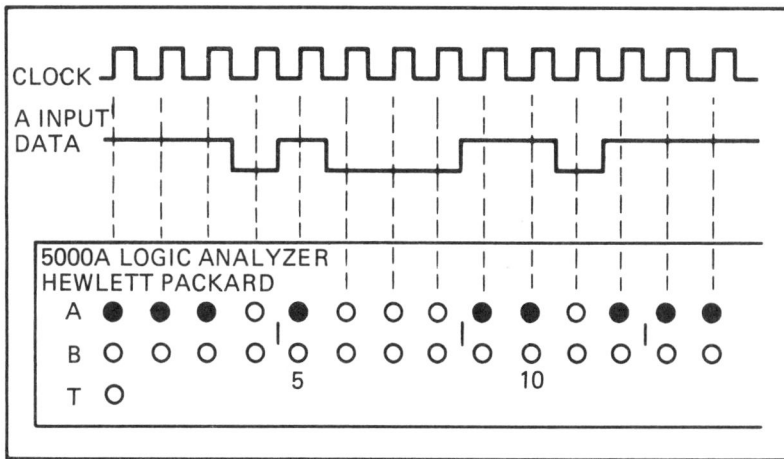

FIGURE 10-21 (Hewlett-Packard Journal)

In addition to the clock input and the A and B inputs there is provision for an external trigger input. Triggering starts or ends a sequence to be displayed. The received data at inputs A and B is stored and is displayed on command. The start or end of the display is determined by the trigger point. At the trigger command either the preceding 32 bits or the following 32 bits can be displayed. Also A and B can be operated in tandem giving a 64-bit sequence from input A.

An important feature of the analyzer is its delay provisions. A desired delay is dialed into front-panel thumb-wheels, the setting of which indicates the number of bits of delay entered and the resulting delay downstream from the trigger point at which the display will start. Delay up to 999,999 bits or 9999 words plus 99 bits can be dialed up. This provision is extremely useful in examining long data streams.

The instrument includes spike detection provisions. In this mode, if more than one logic transition occurs within a single clock period, it is displayed as a spike. If the first transition is positive going, it is displayed as a positive spike and if negative going, as a negative spike (see Figure 10-22).

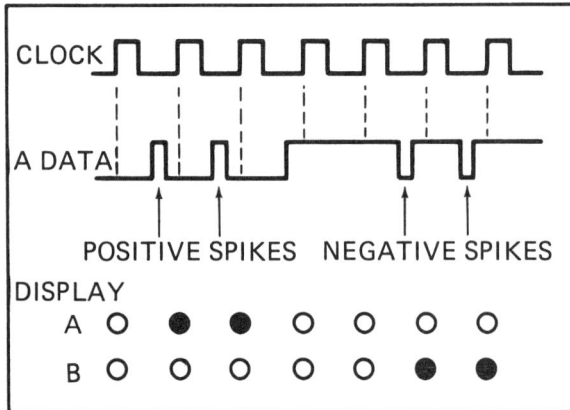

FIGURE 10-22 (Hewlett-Packard Journal)

In addition to the study of disc drives, microprogrammed devices and digital data transmissions, there are many everyday uses for the analyzer. The operation of a flip-flop can be checked for correct division by two by connecting the clock line to the clock input of the flip-flop and the Q output to the input A. Proper division by two will appear on the display as alternate on and off LEDs. AND gates may be checked for proper operation. The number of clock pulses between overflows in a programmable frequency divider can be readily determined. Data bits dropped in a recirculating shift register can be found by comparing the circulating data with a stored reference.

Why logic state analyzers are required in some cases.

Even more powerful tools than the logic analyzers described above are the logic state analyzers. These instead of displaying a limited serial sequence can handle parallel data and, for example, display 15 words of up to 12 bits each on a single cathode ray screen.

The display comprises rows of 0s and 1s as they would appear on a corresponding truth table. As these analyzers accept parallel data, they are provided with 12 input lines. Any lesser number of these lines can be used at any given time. Unused columns are blanked from the display. A clock input provides for using externally generated pulses for strobing data from the input lines into the analyzer's memory.

The following description will be particularly directed to the Hewlett-Packard Model 1601L Logic State Analyzer, a picture of the front panel of which is provided in Figure 10-23.

This analyzer uses a trigger word and can be set to display along with the trigger word (brightened to show its position in the data stream) the preceding 15 words or the following 15 words. With this capability, one can page through a long digital sequence, as long as 100,000 words if need be.

The trigger word is selected by means of 12 front-panel switches each having three positions: HI, LO and OFF. The trigger operates when a word in the incoming data matches the word set-up on the trigger switches.

The triggering and state flow in a BCD counter is illustrated in Figure 10-24. All but the four least significant digits are blanked out. The trigger word is 0000, set in the four switches controlling the four least significant digit inputs. The display from top to bottom shows the trigger word 0000 and the succeeding 15 counts starting with 1, proceeding to 1001 (9), back to 0000 (0) and then to the 15th count 0101 (5).

The analyzer, using its memory, can capture a sequence from a single pass and display it continuously until RESET, which initiates a new acquisition cycle. In the update mode, the display is updated upon receipt of any trigger word. In the free-running mode it initiates a new display cycle each time a display cycle is completed.

Some applications are aided by the addition of an oscilloscope. The analyzer provides a buffered clock output and a pulse each time a trigger word is received. The trigger word can be chosen as the exact digital state around which data or operating conditions are to be checked. For example, a BCD counter is counting incorrectly, resetting on state 89 instead of 99. The analyzer is set to trigger at state 88 which also triggers the oscilloscope. Observation of the master reset bus with the oscilloscope shows a transient at the start of state 90 (see Figure 10-25).

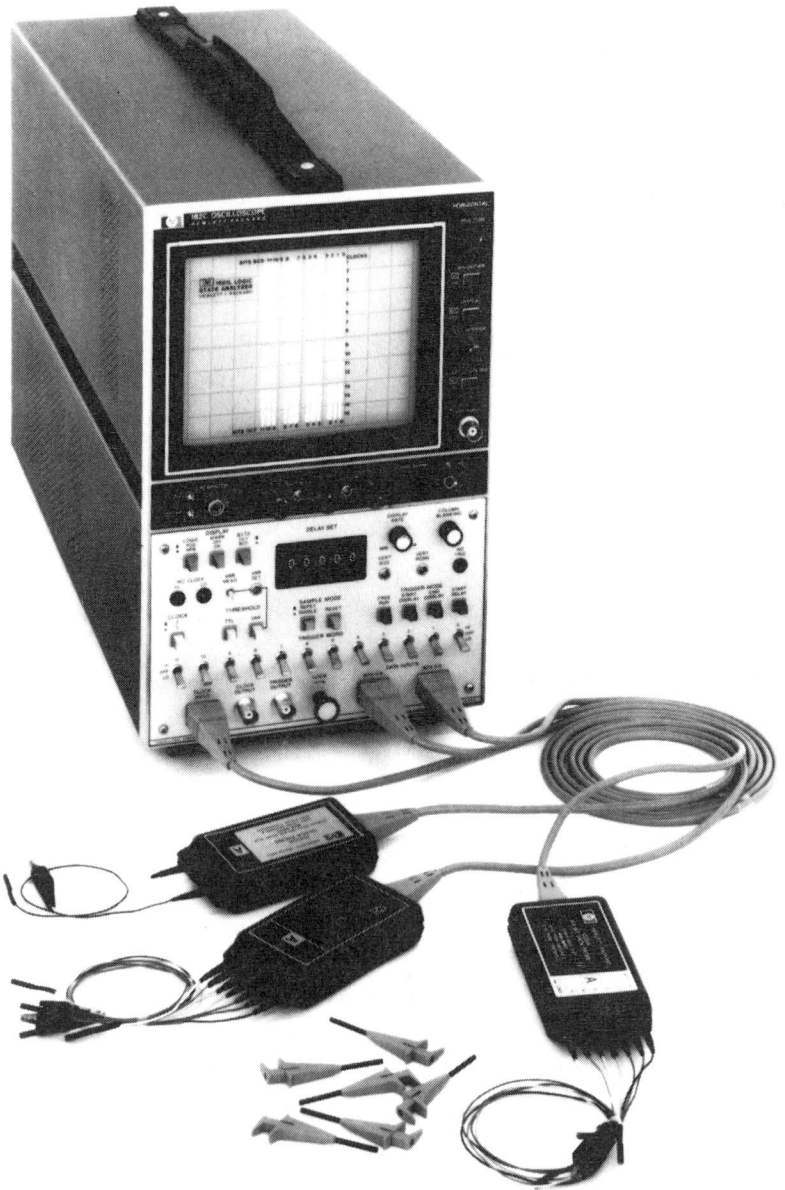

FIGURE 10-23 (Hewlett-Packard Journal)

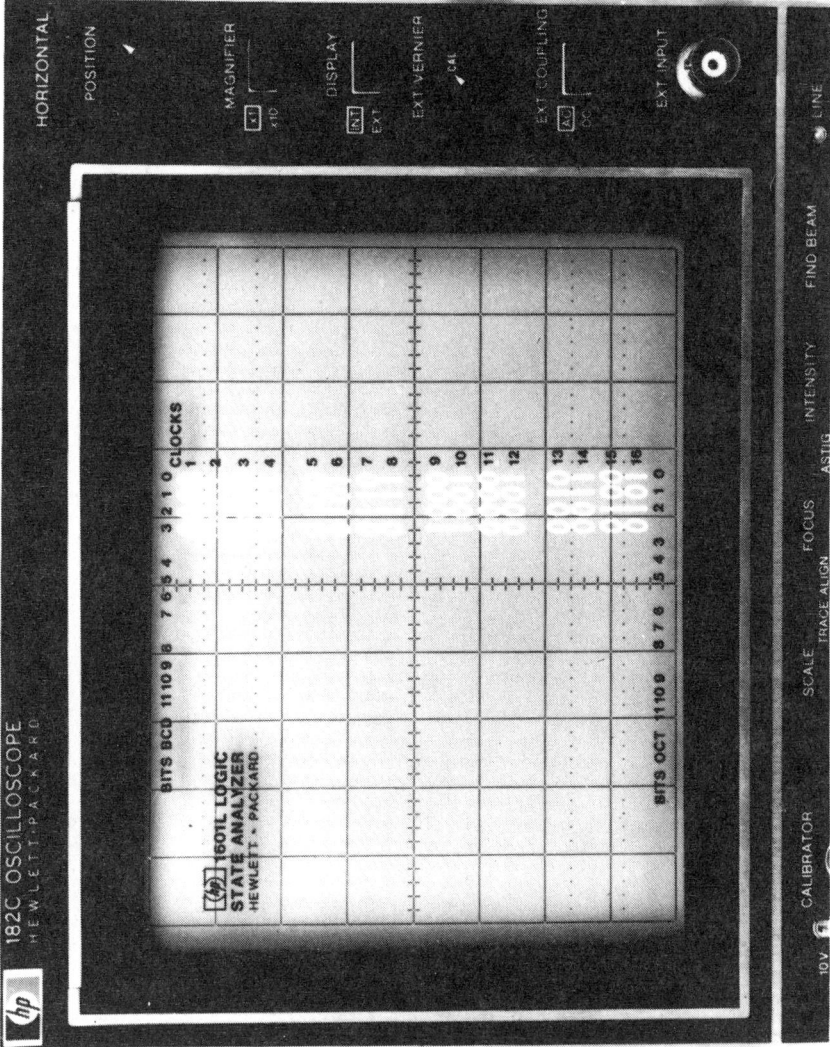

FIGURE 10-24 (Hewlett-Packard Journal)

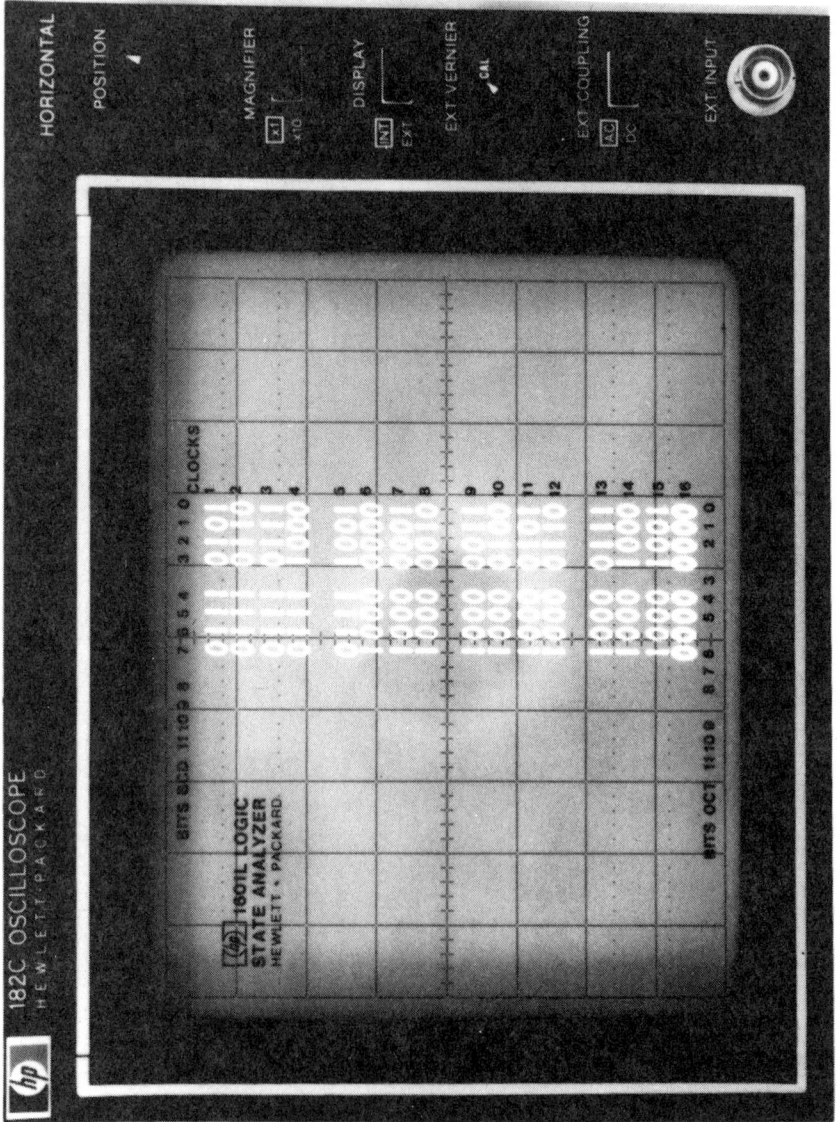

FIGURE 10-25 (Hewlett-Packard Journal)

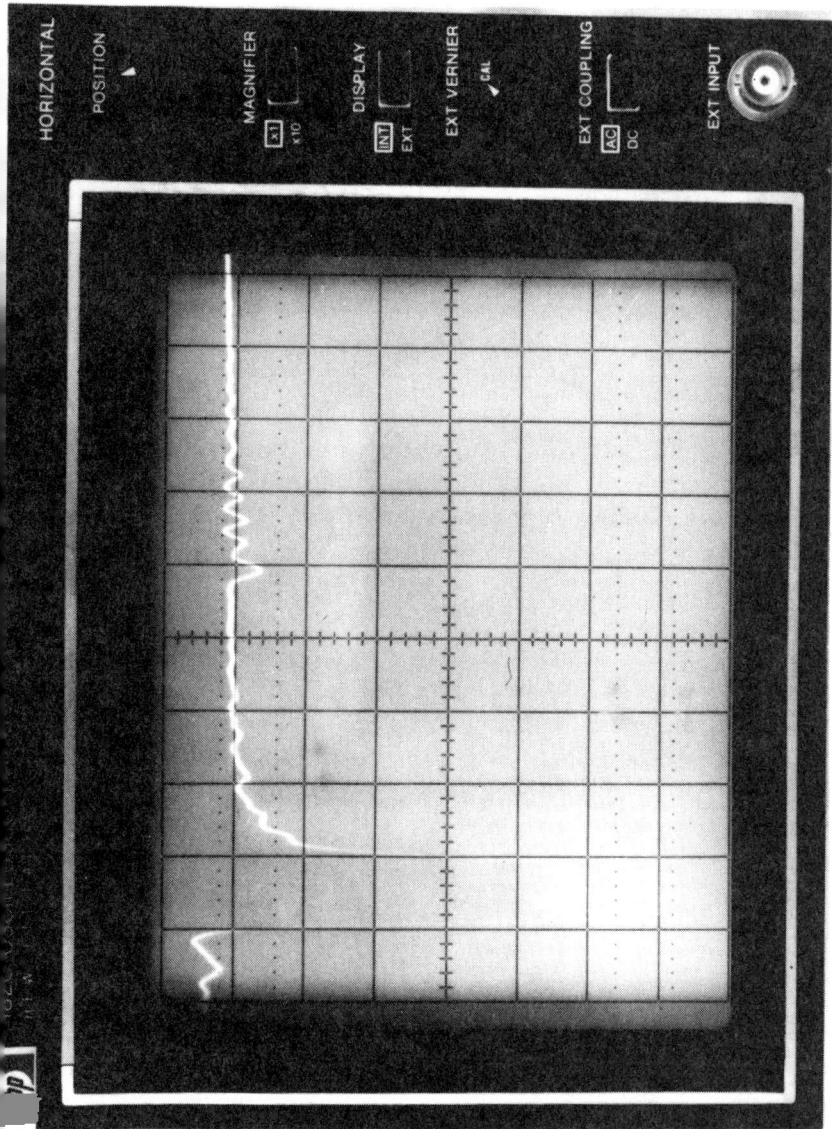

FIGURE 10-25 (cont.)

Uses for the analyzer include examination of "start" sequences, turn-on sequences, calculator-plotter wait loop time, microprocessor operation and, in particular, long sequences of data or digital commands.

Practical information about automatic testers.

Automatic testing had its inception a long time ago. Broadly, automatic testing is any testing to determine characteristics or acceptability of a material, device, circuit or complex system. It is any mechanical or electrical testing that is machine controlled with a minimum of assistance from a human operator. A fully automated system is one which loads the device to be tested, performs complete tests, determines fault location, rejects unacceptable devices, and provides some form of permanent record at each step, including fault locations.

Automatic testing has evolved in the IC field because manual testing of complex ICs is far too slow and expensive to be practical under most conditions. The complexity of many ICs and most circuit boards requires a correspondingly complex test set-up and routine —generally, one which would be completely impractical for a manual test set-up.

There are a number of different test programs or methods used in automatic testing. The choice depends on many factors. A simple method to implement is the random signal comparator where a randomly generated signal is used to exercise the device under test (DUT) and a known good comparison device simultaneously. Failures are indicated when the outputs of the two devices do not match. One disadvantage of this method of test is that since the input signals are random, so are the output signals and no comparison from run-to-run can be made. The test results can be made comparable by using a pseudorandom sequence of input signals. This means that while the input patterns can be considered random as far as the DUT is concerned, they are actually repeated each time in the same sequence so that the output responses are comparable for evaluation.

Another test method may be called "path-sensitizing." Here a program is provided which provides a path through a logic network in such a manner that a logic change along the path causes a change at the output. When the predicted output is not produced, a stuck failure is considered to be detected.

A further test method may be called the interactive test-generation method in which the test system and a test technician act as partners in a closed loop interaction. The technician tells the system under test, "Try this," and the system replies indicating whether the idea worked or not. If it didn't work, the system feeds back information which shows the operator how to resolve the problem.

More complex test systems use automatic test generation (ATG) with computer programs which provide input-stimuli, network digital logic simulation and fault simulation.

The particular method best suited to a given situation depends on the type of circuit being tested, its complexity, the number to be tested and many other factors. Synchronous sytems require different treatment from asynchronous sytems. Complex networks require one type of treatment while sequential networks require quite another.

Automatic test systems (ATS) and automatic test equipment (ATE).

Functionally Ideal System

1. Makes complete functional tests in zero time.
2. Validates all intended functions.
3. Discovers all functional faults in a single pass, including shorts, open circuits, intermittents, and gate stuck at 1 or 0.
4. Identifies location of all discovered faults.
5. Indicates cure for all discovered faults.
6. Automatically handles all input and output of devices, including separation of acceptable from unacceptable devices.
7. Tests any device or system of any complexity and any logic system, even those not designed for easy testing.
8. Easily programmed by standardized programs or by readily generated software.
9. Operable by unskilled operators.
10. Minimum change-over time.
11. Little or no maintenance or down time.
12. Low initial cost and long life expectancy.

Hardware Realities

1. Hardware meeting the requirements of the functionally ideal system does not exist because:

    A. The cost would not be warranted.

    B. The cost would be prohibitive.

    C. By the time an ideal system is produced the art will have evolved into something unexpectedly different.

2. Many sophisticated automatic test equipments are available.

3. The selection of the most viable hardware is a demanding task.

4. Many factors must be weighed in the selection of automatic testing hardware.

5. Both present and foreseeable future needs should be considered in selecting hardware.

6. Advice from reliable and experienced suppliers can be invaluable in hardware selection.

7. Hardware obsolesence can be very expensive.

8. Hardware should be versatile and adaptable to meet rapidly changing requirements.

Software Assistance

1. Automatic test systems for complex devices and systems have often been a joint development in software and hardware.

2. Comprehensive or standardized software or readily generated software is important in a practical system.

3. The cost of the software is often a major item in an automatic test system.

4. An alternative to software is the use of a time shared computer.

Recommendations for users of ATS and ATE.

1. Design for testability.

2. Involve design engineers in the final testing program.

3. Insist on having adequate test points provided for viable testing.

4. Take sufficient time to choose automatic test equipment wisely.

5. Study the various test method philosophies and their application to your products before deciding on ATS and selecting the hardware.

6. Organize testing under control of a data manager.

While ICs are being developed and produced of ever increasing complexity, testing them has developed into an important and challenging allied art. It is not enough to conceive, design and integrate a new device. This device must be tested. This means the device must be testable. Thus, designing for testability has become an important field of endeavor.

11

How to Use ICs
in Hobby Projects

ICs in the hobby field add potential and interest.

Popular electronic magazines have been featuring integrated circuit hobby devices for several years. There is no point in trying to compete on their terms, which is to describe the assembly and operation of predesigned kits or similar projects. There is a certain amount of satisfaction to be derived from building these circuits. However, I propose a different approach. I propose to describe a case history of an integrated circuit device, starting with a search for a field needing new ideas; deciding on such a field; exploring the possibilities and coming up with an idea for a really new device; outlining the desired functions; laying out a combination of circuit devices to carry out these functions; exploring alternate ways to carry out the functions; making tentative tests on various components; revising the lay-out in view of experiments; building a model to test overall operation; simplifying the model; testing performance; debugging; when satisfied with the operation of the model, finalizing the design with schematic diagrams, wiring diagrams and component list and specification. The overall result is an object lesson in conception, analysis, construction and testing of an integrated circuit device. It will demonstrate the value of an understanding in depth of individual components, their interactions and practical construction.

Before this project is described in detail some general advice will be addressed to hobbyists. The hobby fields are many and varied, appealing in different ways to different people.

Hobby builders are generally more familiar with analog projects than they are with digital. As a matter of fact a number of books have been written on IC projects which are entirely devoid of any mention of digital ICs. However, digital ICs have many applications in the hobby field. The special bibliography at the end of the chapter lists a number of these projects which are partly or completely digitally activated.

Digital clocks are a "natural." Two approaches are possible. One is to use a crystal controlled oscillator of such a frequency as to be readily divided to provide seconds, minutes and hours; the other is to use the 60-cycle ac power line as the reference frequency. Of course, the former is the only one which will operate on batteries. Many variations of the integrated circuit clock are possible, particularly in the manner in which the time is displayed.

Tips on building hobby projects.

The hobbyist may derive satisfaction and pleasure from the entire process of planning a project, experimenting with various components, building and perfecting a finished model and using the model. This may involve him in a number of different fields, from electronic circuitry to musical composition. It can be particularly fascinating if one has a great deal of time to spend pursuing this hobby.

Others may derive their greatest pleasure from the electronics involved in planning and executing a project. Still others are mainly interested in the end result, doing the building mainly to save money on the final product.

Before starting a project, examine the following factors:

1. Do you have sufficient information about what you may be able to generate by experimentation to complete the project with a good chance of success?
2. Are the component parts reasonably available and at prices you are willing to pay?
3. Do you have the necessary tools to carry out the project in a workmanship manner?
4. Do you have the time to use and enjoy in carrying out the project?

5. Don't be fooled by kits "anyone can build." The question is not only can you build it but also can you build it well and can you get it to work?

6. Do you have a proper place to build your project, to experiment or test it and to get it working?

A great deal of the enjoyment and learning experience which can accompany building hobby projects is lost if one does not understand what is going on in the circuits one puts together. Test equipment provides a means for increasing understanding, checking for correct wiring, locating malfunctions, and obtaining optimum operation of the finished device. A low-voltage, low-current ohmmeter is the most basic and versatile of all test instruments. Since they are readily available as multimeters, a dc and ac volt and current meter should be included. For digital IC work, a simple logic state probe or indicator would come next on the list of importance. More sophisticated tests require signal sources in the form of pulse generators, audio and radio frequency signal generators and, most useful of all, a cathode ray oscilloscope. In order to calibrate the sound level meter described later in the chapter, a sound level meter is required.

Now, going back to discuss these points in more detail. There are many pitfalls in hobby project information. The *Popular Magazine* description of a project may assume you have more experience or ability than you actually have. It depends a great deal on the writer and to whom he assumes he is writing. Don't overlook the sad fact that many magazine descriptions contain errors, and if you can't spot them and correct them you may be stymied. Magazine articles are necessarily abbreviated, often leaving out theory and trouble-shooting information. Kits are not always foolproof, either. A kit made in Japan with assembly and wiring instructions translated from the Japanese can be frustrating. Some foreign-made kits even use unfamiliar electrical symbols, presenting a further problem. Remember that proficiency in almost anything comes only with experience so start with something simple and progress. Learn by doing.

Before becoming committed, make sure *all* the parts required for your project are available and at prices you are willing to pay. Just the fact that manufacturer and type numbers are specified does not guarantee this. An engineer working for a large company can often pick up sample parts sent to his company for approval but not readily available

to others. Many manufacturers sell direct or through wholesale chan-
nels which do not cater to the individual. When you need a 50¢ part
and find it is only available from a supplier with a $25.00 minimum
charge you may find the source impractical. There are a number of
mail order dealers who sell surplus components. You cannot assume
that their quality control is sufficient to guarantee perfect parts. If you
are planning to do a substantial amount of project building you may
find it very helpful to set up your own test facilities. Most testing of
integrated circuits can be accomplished with reasonably simple equipment.

The tools which will make IC projects easy include a fine-tip
soldering iron, fine rosin core solder (0.05 in diam. or less), a small
vise, a hand magnifying glass or loop of 3 to 5 power, a desoldering
iron (in case you make a mistake), small needle-nose pliers, small
wire-cutting diagonal pliers, wire strippers, solid tinned copper wire,
24-26 gauge and fine spaghetti to match.

A rather basic decision in using integrated circuits is whether to
use sockets or experimental type printed circuit boards. For performing
many different experiments and where a final permanent device is not
the object, sockets mounted on Bakelite box tops with numbered pin
tip jacks are very useful. For a finished product, printed circuit type
boards are available in various sizes with edge connectors. These make
really professional looking devices.

Integrated circuits are great time savers. Individual circuits are
standardized so that all one has to do is to assemble functional blocks.
Still, electronic projects do take a substantial amount of time. Some kit
manufacturers state a recommended assembly time. Only very simple
projects can be completed in an evening or even a weekend. To enjoy
the study, assembly wiring and testing, be sure you have ample time to
complete it without rushing. Be prepared to spend as much time debugging your project as you took to build it. A simple mistake in
wiring, an unsoldered joint or some other seemingly small error will
prevent proper operation and may take some real time to locate.

Don't assume, just because it is a ''kit,'' that anyone can build it
and that it will work without fail. Some of the problems have been
mentioned above. An electronic circuit is not like an automobile. If a
spark plug is shorted the symptom of a misfiring of the engine is pretty
sure to give it away—but one error in an electronic circuit and nothing
may work and there are no symptoms! Kits can have missing parts.

Kits can have improperly identified parts. Kits can have insufficient directions for *you* to build it readily. In some cases kits are incomplete; some special integrate circuit is supplied but no keyboard, or readout, or other adjunct to a complete device. Make sure you know what you are getting and whether or not you will need further parts to complete, and look for some assurance you will succeed in building what you intend.

Having a proper place to build your project, test it and perfect it can be important to the real enjoyment of hobby projects. It should be a place dedicated to your hobby where you can come and go as the spirit moves and time permits. This requirement is at least partially dictated by the fact that most projects cannot be completed at one sitting. Be sure you have good light, ac power and a table or bench which will not be damaged by melted solder or a misplaced soldering iron. Have enough room to spread out your diagrams, parts, test equipment and tools so that they are conveniently at hand but uncrowded.

Start with a simple project. Learn to make a quick, neat and secure soldered joint without causing the solder to over-run the next printed circuit pad. Learn to recognize a good soldered joint in contrast to the so-called rosin joint where the rosin holds things mechanically but insulates them electrically.

Before applying power to your hobby circuit check it for the following errors:

1. Has an incorrectly drawn circuit been slavishly copied on the assumption that it was right? Reason out the signal flow through the circuit and if it doesn't look right, check very carefully. If necessary, try out part of the circuit to find out what is correct.
2. Incorrect wiring. If possible, have someone else check your work. If not, call out point-to-point connections and have someone check you, or check the circuit yourself point by point. Check continuity with an ohmmeter on a high resistance scale so as not to damage any transistors with high current. Check any diodes for correct polarity.
3. Wrong pins in the socket. Make sure pin 1 of all modules is in socket pin 1, and so on.
4. Wrong bias polarity. No matter what the circuit shows, be sure the bias polarity is correct.

5. Check all soldered joints mechanically, electrically for low resistance continuity and visually with a magnifying glass.

One question which may be ambiguous is what to do with unused inputs to gates and flip-flops. This is a matter of reason and/or the manufacturer's statements. Some logic families see 0 at an open input while others see 1. What do you want the unused inputs to see? If in doubt as to what will happen if left open, connect unused inputs to the appropriate 0 or 1 bias point.

Be sure your circuit bias voltage or voltages are within specification. In a battery operated device it will be important to check the battery voltage from time to time. Low voltage will often result in malfunction due to falling drive voltage and current from one module to the next. This situation is particularly likely to happen where several input circuits are being driven from one output circuit. High voltage can cause overheating and eventual failure. Power dissipation, it must be remembered, goes up as the square of the increase in voltage.

How to invent a new hobby device from start to finish.

The following will trace an invention from the initial concept through reduction to practice and improvement and simplification to the point of providing the basis for a manufacturing prototype.

Someone had suggested the subject of electric trains. Here was a growing field and one which seemed to have failed to utilize fully what electronics has to offer. What could be done with electric trains that would be new and interesting and that would involve the use of integrated circuits? After some reflection the idea presented itself—"How about a voice control system for electric trains?" As far as could be determined nothing had been done along this line.

Starting to analyze the problem and how to implement a voice control led to the consideration of how the human voice can be used to initiate commands. There are several possibilities. The use of a sytem which analyzes voice frequencies or matches spoken words or the like seemed very complex to carry out and expensive beyond its real value as a hobby item. The idea which seemed promising because of its simplicity was to convert voice syllables into pulses and to decode these pulses.

The simple set-up shown in Figure 11-1 proved that rather definite pulses of substantially equal length could be produced with a simple microphone, amplifier and rectifier charging a capacitor. Speaking deliberately produced substantially equally spaced pulses of approximately one second duration. The set-up involving three simple commands and the resulting one, two and three pulses was arrived at after some experimentation. The commands finally chosen were "stop" (one), "back up" (two), and "go ahead" (three).

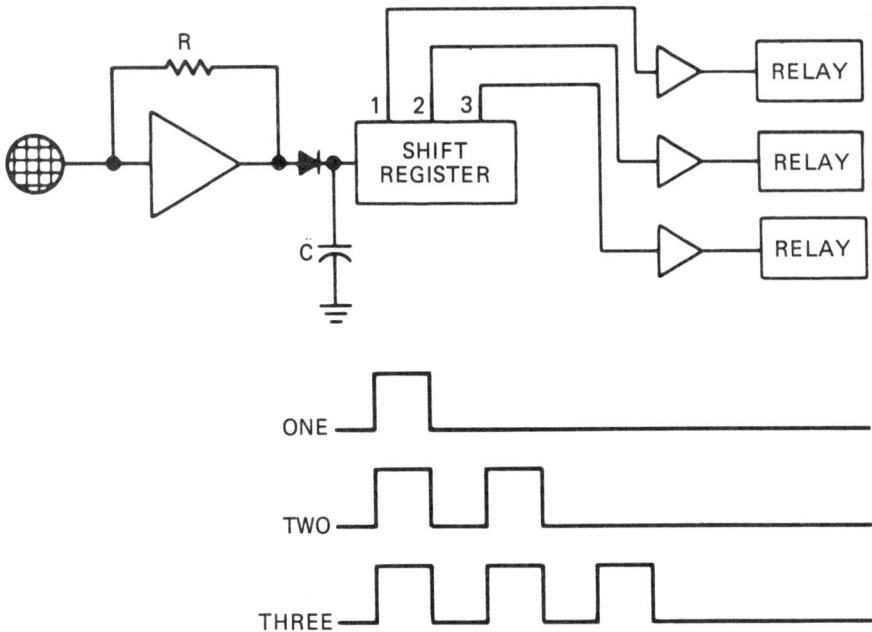

FIGURE 11-1

The amplifier was a standard operational amplifier which has an offset. It was found that if the offset was such as to bottom the amplifier output in the positive direction in the absence of any input from the microphone, the negative going pulses were substantially equal to the positive bias plus the negative bias on the amplifier. However, some amplifiers showed an offset in one direction and others an offset in the opposite direction. It was thus necessary to provide a bias to insure an initial bias in the desired direction. A resistor from the negative bias voltage to the high side of the microphone as shown in Figure 11-2 was simple and effective. Providing some excess bias current reduced the sensitivity of the microphone input so that the commands would not be triggered by ambient noise but only by close talking.

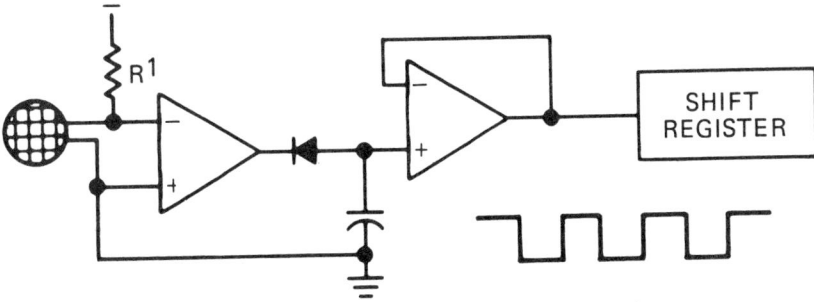

FIGURE 11-2

Next, a simple block diagram for further implementation was drawn as shown in Figure 11-3. At this point further ideas had to be generated. How were the pulses to be decoded to provide the desired commands? The easily available functions with simple, inexpensive integrated circuits include operational amplifiers, gates, flip-flops, counters, shift registers and timers.

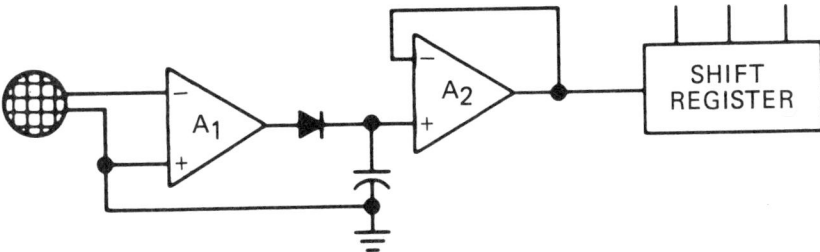

FIGURE 11-3

First, experiments were performed with the 555 type timer (Figure 11-4), since this is a versatile device, operable as a monostable or as an astable multivibrator. It could be used for two functions: first, to generate an output inhibiting pulse, and second, to generate pulses to fill a shift register before counting input pulses. It seemed necessary to inhibit the output at the start of a sequence of commands until the commands are complete. Otherwise, the output relays will chatter away as the command develops, which is obviously undesirable. Also, if a shift register is to be used to decode the commands, it must be in some predetermined condition at the start of the commands. One way to do this is to trigger an astable multivibrator at the start of the commands, detect when a circulating 1 in the shift register is in position 4, stop the astable multivibrator and count the number of syllables in the command in the shift register. For example, the two-syllable command "Back up" would place the 1 in position 2. When the monostable returns high, opening the output circuits, the number 2 relay ("back up" circuit) would be energized.

One of the practical problems which developed with this circuit was due to the possibility of the circulating 1 becoming lost or even being multiplied. If the 1 disappeared the system ceased to function. If a second 1 appeared, the output commands become ambiguous.

The next step was to clear the shift register at the start of each command and then to set in a 1 in the first stage while using 2, 3 and 4 as the useful outputs. The timing diagram shown in Figure 11-5 was drawn. The "Clear" and "Set in a 1" can be monostable multivibrators. A block diagram shown in Figure 11-6 was drawn to implement the functions of the timing diagram.

At this point a problem, common to the experimenter, developed. Three more 555 timers were not readily available. However, at this point no astable multivibrator was being used so a switch was made to the readily available 74121. This is intended for use as a monostable multivibrator, and it operated well. The Q and $\overline{Q}$ outputs and the multiple inputs available made a very flexible building block. Some experiments were performed with the 74121 in order to become familiar with its operation and to determine the most compatible external components. It was found that apparently the best operation on long periods was obtained with an external resistor no greater than 10K and with whatever capacitor was required to provide the time period.

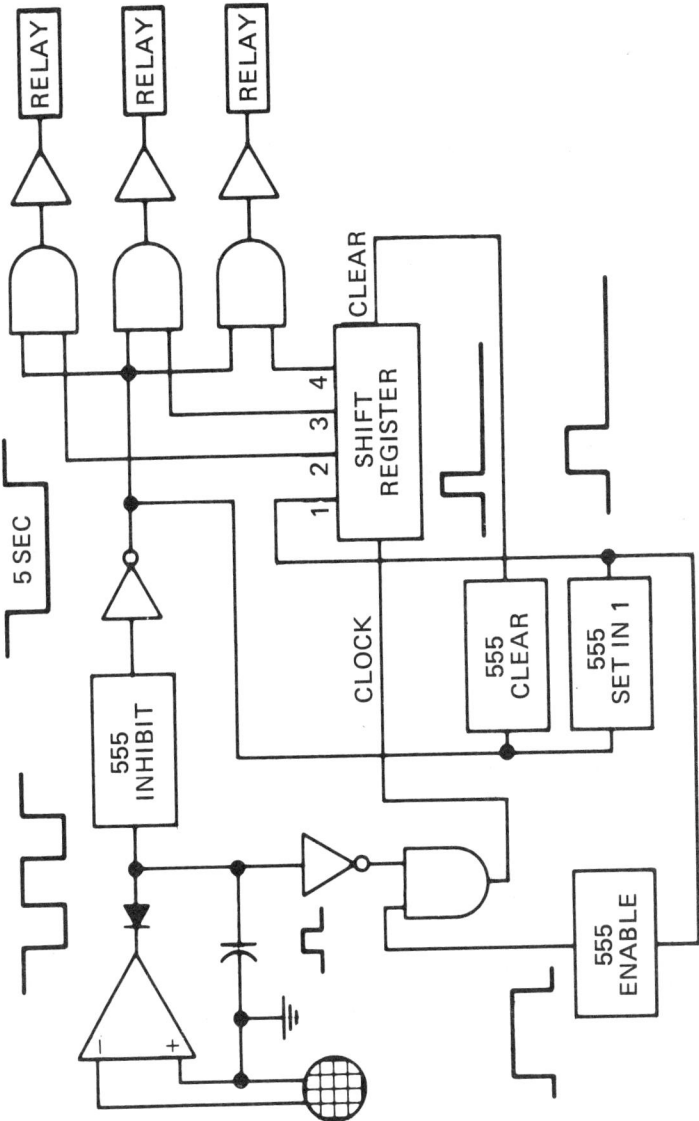

FIGURE 11-4

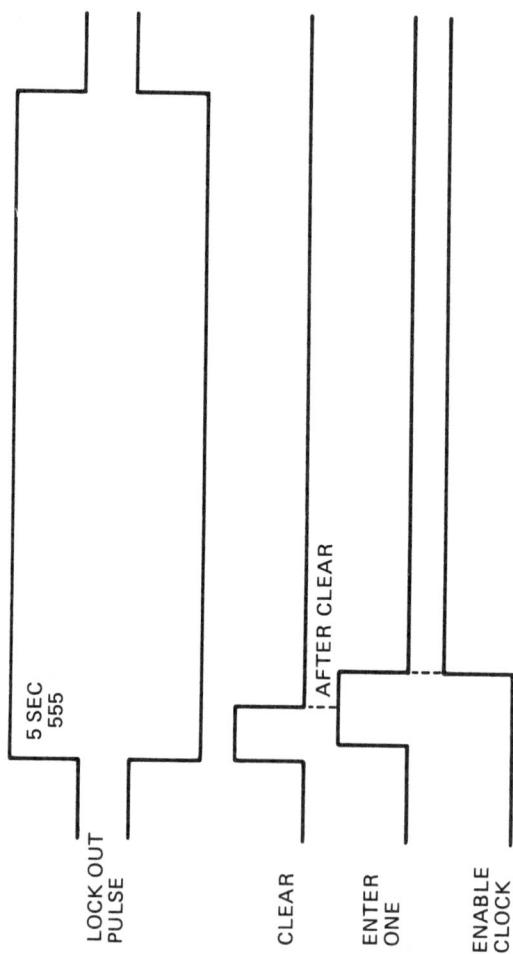

FIGURE 11-5

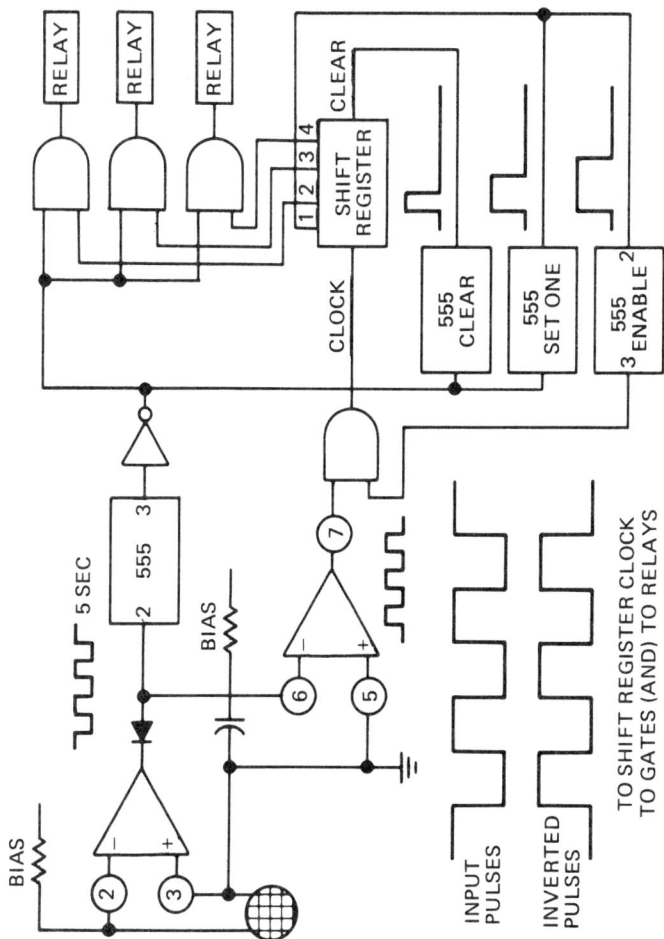

FIGURE 11-6

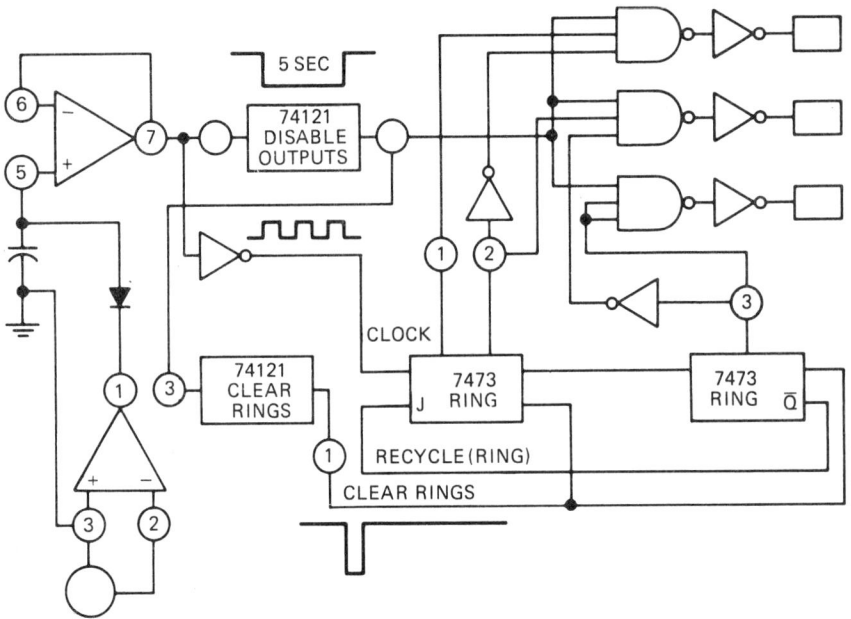

FIGURE 11-7

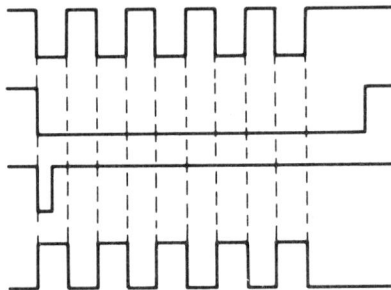

THIS PERMITS OMITTING
THE 7402 NORS—

FIGURE 11-7B

At the same time it was decided to make a shift register from 7473 J-K flip-flops, two of which (each containing two flip-flops) would provide a count of 4.

(It should be noted here that there has been and is to be some simplification in this presentation over what actually was tried. Some experimenters, no doubt, will do as well or maybe even better, while others will do worse.)

The use of the 7473 flip-flops as a shift register had its advantages and its disadvantages. One advantage was that with the "Ring" connection shown in Figure 11-7 the "Set in 1" function was not required since at the end of "Clear" the last $\overline{Q}$ is 1, and the clock pulses advance this 1 sequentially to outputs 1, 2 and 3. The disadvantage is that decoding is required among the outputs since the 1 is not only advanced from 1, to 2, etc., but it is followed by a second and third 1. Thus, at a count of three, outputs appear on outputs 1, 2 and 3. On the other hand, decoding is not difficult. Figure 11-8 shows how 1, 1 + 2 and 1 + 2 + 3 can be decoded into 1 or 2 or 3 exclusively.

When the circuit of Figure 11-8 was set up and tried out, it was found that the syllable induced pulses were ragged and attenuated. This was traced to the fact that the inputs to the devices connected across the

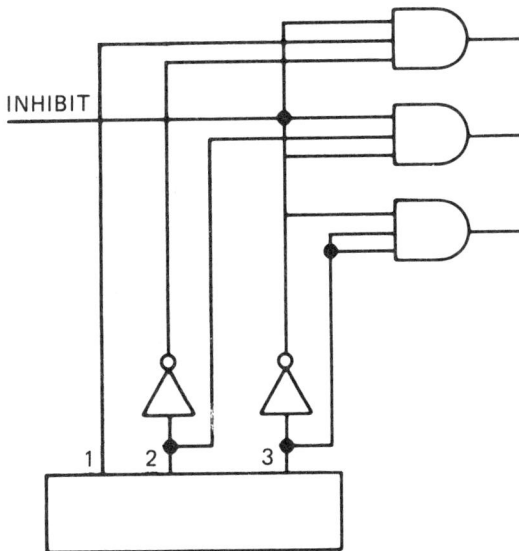

FIGURE 11-8

filter capacitor following the rectifier were loading it and speeding its discharge. This problem was met by using a high input voltage follower after the capacitor, as shown in Figure 11-9.

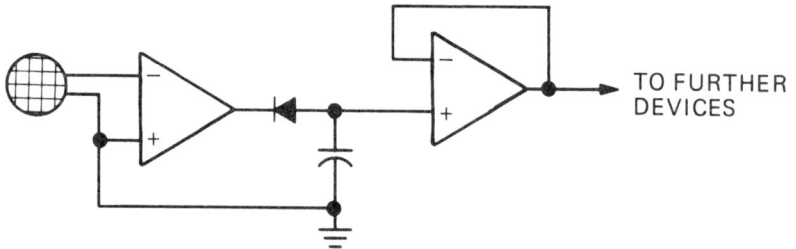

FIGURE 11-9

Finally the circuit shown in Figure 11-10 was evolved. It was assembled and wired on an experimenter's type printed circuit board. The functioning of the circuit can be described as follows: Speech commands are amplified, rectified and integrated to provide one, two or three negative going pulses of approximately one-second duration each (depending on the skill of the command operator); these pulses are of maximum amplitude, i.e., from positive output saturation to negative output saturation provided by an initial offset bias; the pulses are amplified by a high impedance follower in order not to load the pulse-producing capacitor and in order to be able to drive reasonable load devices; following is a monostable multivibrator connected to generate an output pulse of the order of 5 seconds for disabling the output circuits during the input command interval; the leading edge of the disabling pulse triggers a monostable multivibrator generating a sharp pulse which is connected to clear the shift register ring; this results in a 1 at the $\overline{Q}$ output of the ring which is also connected to the ring input; this 1 is entered at the 1 position of the ring by the first input responsive pulse, inverted and applied to the clock input; each succeeding pulse clocks an additional pulse into the register; at the end of the command there will be a count of 1, 2 or 3 in the register; the disabled pulse to the output NAND gates ends; the count in the shift register is decoded: a 1 if the sole count is decoded in the first NAND gate where the 1 output is applied directly and the 2 is applied inverted since the 1 is not the true output if an output at 2 also exists; similarly 2 is

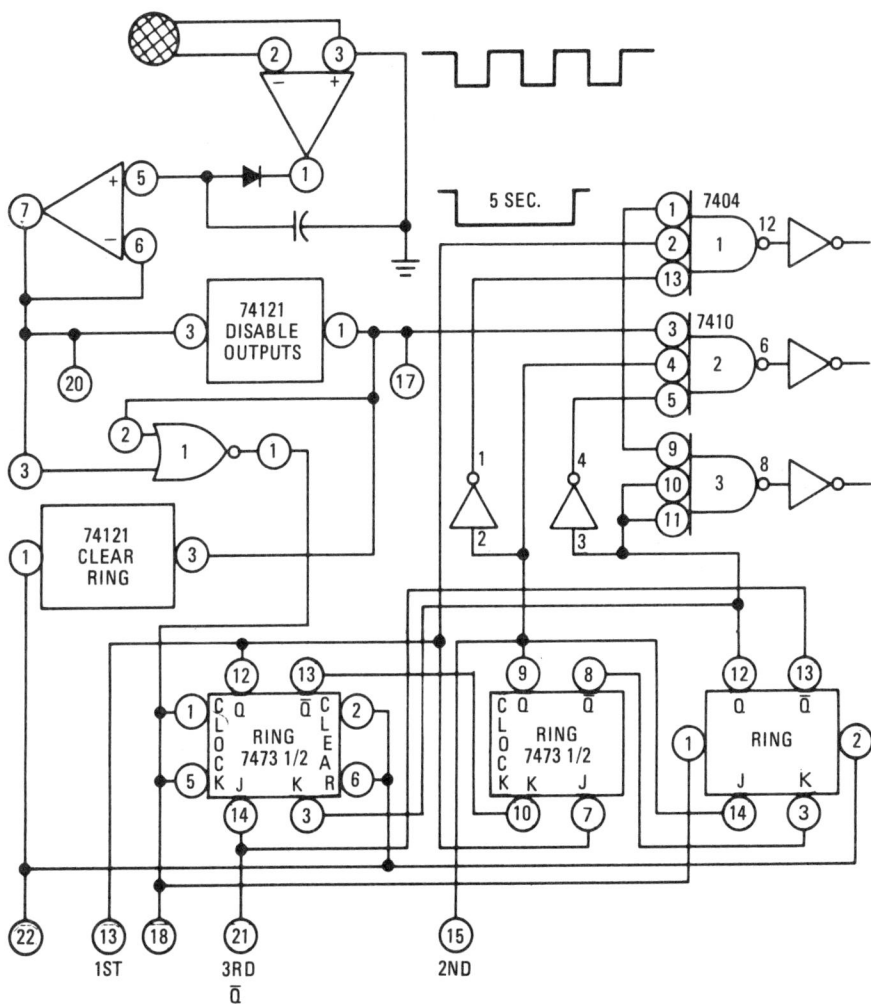

FIGURE 11-10

PULSE	Q1	$\overline{Q2}$	Q2	$\overline{Q2}$	Q3	$\overline{Q3}$
0	0	1	0	1	0	1
1	1	0	0	1	0	1
2	1	0	1	0	0	1
3	1	0	1	0	1	0
4	0	1	1	0	1	0

PULSE	STAGE	Q	$\overline{Q}$
1	1	0	1
1	1	1	0
2	2	1	0

applied directly to the second gate with the 3 output inverted; and the 3 output is applied directly since, if an output exists at 3, it is the true output; the NAND gate outputs are inverted and the resulting positive signals energize the corresponding circuit control relays.

Now there may be possible further simplifications and there may be better solutions to the problem from quite a different angle. But while it is never certain that one has found the best solution, there is a practical limit to the amount of time and effort one can expend on a practical problem. If the final solution seems to be a good one, meets practical criteria as far as reliability, cost, etc., are concerned, one can accept it—at least for the time being. Experience in time may indicate that the problem should be reworked. Competition may force reconsiderations. New components may point to a new approach. If the item turns out to be one that will sell in large quantities, redesign may be required to provide a more economical design.

The matter of patenting is also one to be considered from several angles. First of all, is the result of such innovative quality as to be readily patented? Further, is the potential business worth protecting to the tune of $1000 or more to be invested in a patent, to say nothing of possibly astronomical expenses in case of infringement prosecution?

Sound level indicator is another unique project described.

An interesting and useful instrument employing both linear and digital integrated circuits is the sound level indicator described below. This indicator, by using a series of colored lights to indicate discrete sound levels, does away with range switching and the expensive indicating meter. The colored lights are more meaningful if they logically represent predetermined sound levels; for example, green set to go on at 40db shows a very quiet environment, blue set at 60db shows a low normal environment, white at 80db a high normal, while yellow at 100db is a warning level and red at 120db shows the sound level has reached a danger level.

Advantages of this system are:

1. Fast response not complicated by the dynamics of a meter response.
2. Visible at a distance (when placed at the back of the room one can at once determine whether or not his voice is carrying to the last row).

3. Shows how much desired sounds are above the ambient noise.
4. Stable amplifying technique using operational amplifiers seldom requires calibration.
5. Many uses possible due to non-technical level indication.

The sound level indicator circuit is shown in block form in Figure 11-11. The signal generated by the sound picked up by the microphone

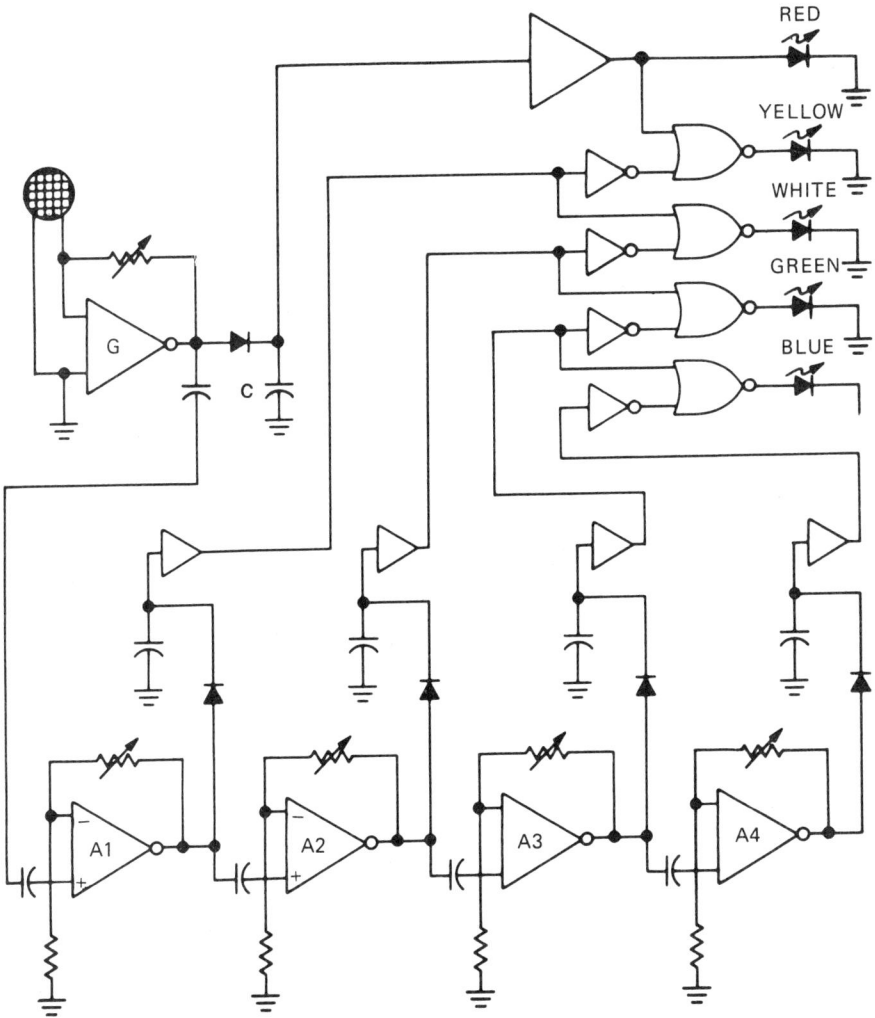

FIGURE 11-11

is amplified in the first amplifier sufficiently to produce a rectified voltage across capacitor C which has an amplitude to just turn on the red light (LED or lamp with red jewel) when the sound level at the microphone just exceeds 120db. (Other maximum levels may of course be chosen.) Each succeeding cascaded amplifier is provided with its feedback set so that the gain of each stage is exactly 10 times or 20db. Turning to the last stage, the signal has been amplified 80db above the level necessary to trigger the first stage lamp and hence a sound level of 120-80 or 40db will trigger the last stage. In order to provide an exclusive output to the lowest level lamp (blue), the output of the last stage is inverted and NORed with the output of the next to the last amplifier output. Only when the last stage is at 1 and the next to the last stage is at 0 will the NOR gate provide a 1 to the blue light source. A similar decoding scheme is used for the next higher output (20db above the last stage); namely, the stage to be indicated is inverted and NORed with the next higher output stage. This second step lights the green (60db) light. The decoding is continued in the same manner so that only one light is on at a time and this light indicates a sound level between its predetermined setting and the next higher level. The final stage from right to left, being the highest level stage, needs no decoding since it can only go on at the maximum level or above and all other lights will have been gated off.

Music synthesizers make use of ICs.

Music means many different things to different people and even to the same people at different times. Familiar combinations of tone, tempo and melody as well as other characteristics generally mean acceptable music. Conversely, unfamiliar components can readily mean unacceptable "music." From the standpoint of the average American, the melodious productions emanating in familiar form from familiar instruments such as violins, cornets, organs, pianos and so on are considered pleasing music. What then is the role of music synthesizers? Are they to imitate conventional instruments, or are they to innovate unfamiliar sounds and combinations? Both are probably viable and hence music synthesizers must be divided into two distinct classes: first, those which imitate conventional instruments in conventional combinations of tempo, melody and other characteristics; and, second, those which create new hitherto unknown tones, combinations and tempos.

Still further, the method of playing may be by means of conventional keyboard, or by completely new means suggested by digital technique, or by hybrid means. With all of these and more possibilities, the description and characterization of music synthesizers becomes indeed complex. To further complicate the matter, magnetic recording may be used to provide an intermediate or an end result in the form of a recording, which is, however, not a real time performance device. Since the possibilities provided are almost unlimited, the present discussion will be aimed at the individual means, leaving the final combinations up to the reader's imagination.

A number of widely available integrated circuits are "naturals" for music synthesizers. First of all, music requires tone sources. At least two—the voltage controlled oscillator (VCO) and the astable multivibrator—are the most common. Both are capable of being remotely controlled (as by keyboard or toggle switch), are capable of producing tones over a wide frequency range, and are susceptible to waveform variety. The astable multivibrator at sub-audible frequencies also provides a tempo generator for modulating the audible tones. Figure 11-12 is a block diagram of a simple tone and tempo generator.

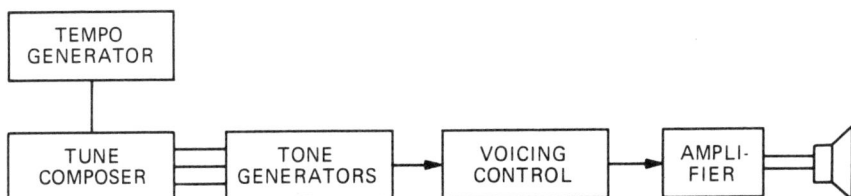

FIGURE 11-12

One of the unique features of music synthesizers is the tone-modifying capabilities provided by integrated circuit techniques. Conventional music is tied rather closely to the available instruments. Violins are violins, trumpets are trumpets; and, while conventional instruments may be plucked instead of bowed, muted, etc., the variations are still finite and have been explored by many musicians over the ages. Music synthesizers open a whole new and an almost unlimited range of permutations and combinations. This, however, poses one of the problems with such music—how does one create really new sounds which are pleasing or at least acceptable to people long accustomed to conventional music? Perhaps the problem is somewhat mitigated by many far-out creations of our day.

A number of other integrated circuit functions provide practically all of the other desirable attributes to music. Timers, shift registers and ring counters provide tempo, tone sequencing and repetition. Control can be by means of toggle switches or programs can be placed in memories. Prepared programs can be placed in read only memories (ROM) and called on by means of simple programs. Switching is provided by optocouplers so that a system can be devised which is programmable with a bare minimum of mechanical switching. However, real time playing or composing can be carried out with a keyboard or groups of toggle switches or both. See Figure 11-13.

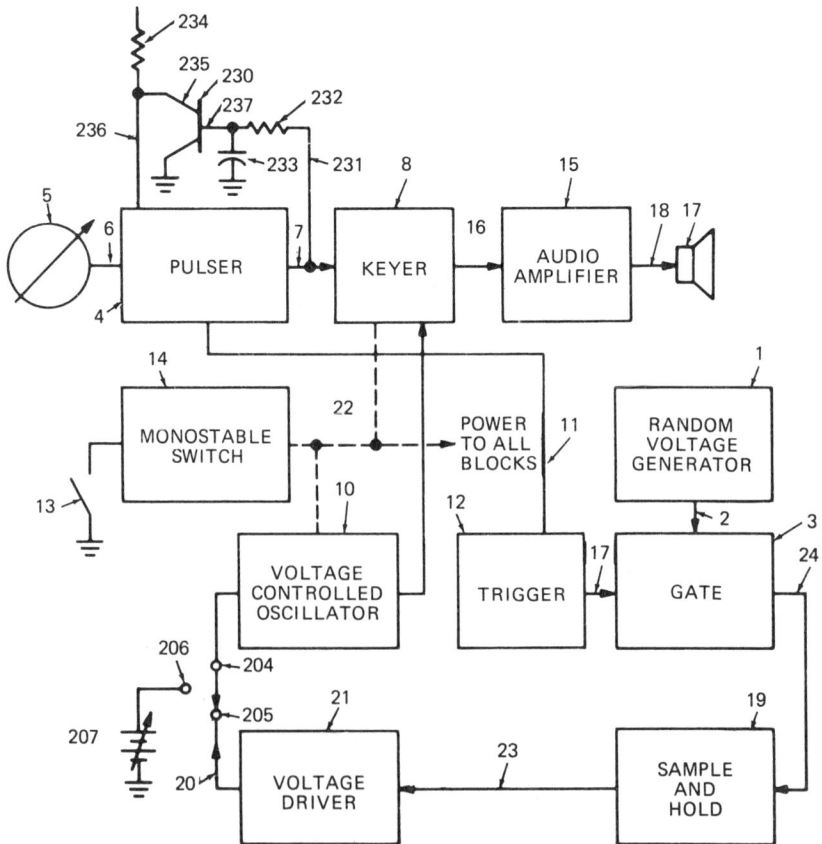

FIGURE 11-13

Still another approach is the digitizer type of operation where music and speech are digitized, i.e., converted to digital form and analyzed. To reproduce the music or speech, the process is reversed and the digital representations are converted to analog voltages and reproduced. This mode of operation can be computerized and hence may have applications where complete mechanization is required. This computerized speech and music, however, is aimed more at computerized reproduction of known words and phrases than at original composition. The creative process suffers in the meantime. See Figure 11-14.

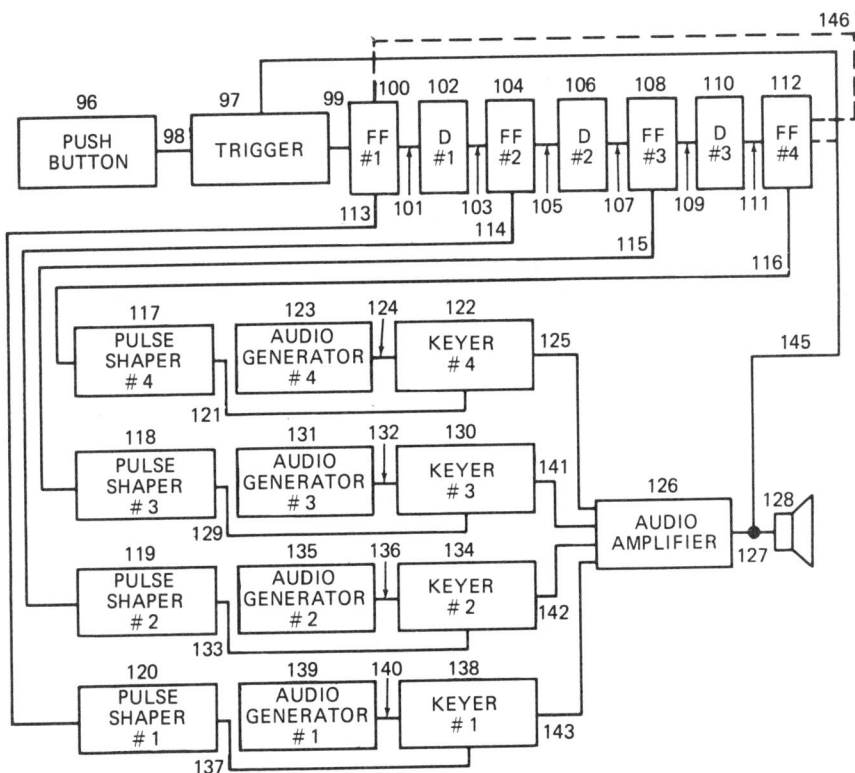

FIGURE 11-14

The Moog Synthesizer has been rather widely used. This sytem is played by means of a keyboard and a great multiplicity of switches. In this way, it is based on the organ technique of keyboard with many tone-modifying stops. However, the electronic modifications are capable of providing a much wider variety than has been possible with any pipe or electronic organ. This is at least in some measure due to the creativity of Robert Moog and his knowledge of electronics. The Moog devices are complex and expensive and generally out of reach of the hobbyist.

There are many other systems available either as kits or as complete instruments. The keyboard instruments are generally more expensive and possibly less exciting since they are essentially extensions of conventional instruments such as the manual organ. Other devices range from very simple one-octave devices played with simple key switches to very elaborate devices using switches, memories and other means. One trying to decide on what type of system to choose should consider the purpose. If it is to provide electronic music of a more or less conventional nature, a basic keyboard device is the answer. If it is to explore new sounds and new creative processes, the more complex switch and memory or similar device should be sought. As a general thing, the ability to create new sounds has gone far ahead of the creative efforts to put together new acceptable "music" or whatever one wishes to call the new synthetic sound patterns.

For those interested in getting involved in the electronic music field, a very informative article entitled "Introduction to Electronic Music" by Don Lancaster was published in *Popular Electronics*, Oct. 1973, pp. 35-37 inclusive. At the end of this chapter there is a bibliography of some of the most revealing articles on the subject, including short summaries of the contents.

It is only natural that electronic organs are being built with more and more integrated circuits. The tone generators, tone modifiers and other attributes to electronic organ circuits are widely available as integrated circuits. The separation in philosophy and construction and the result between the Moog Synthesizer and the modern electronic organ is rapidly getting smaller.

Many other interesting projects have been devised using ICs.

There are many other interesting hobby projects available using ICs. The IC greatly reduces the time required to complete a given project and makes many otherwise too complex projects completely practical. The all-electronic clock, electronic watch, pocket calculator and many others fall within this category. Manufacturers of ICs often can supply hobby instruction sheets or booklets. Popular electronics magazines run them almost continually.

There is one very unhappy note in all this, however. That is that many ICs are classified and sold as hobby units with the statement, "This is an untested chip and some function will not operate." This function could spoil all the fun. To make matters worse, instruction sheets furnished with these and other chips are often quite inadequate for the hobby builder to complete his project. Again the magazine articles are generally quite lacking in theory or mode of operation, which should be of prime interest in the hobby field. There are also many books which are too elementary to create any real or lasting interest. Much could be done to correct these conditions and to make the hobby field both interesting and informative.

Special Bibliography of Hobby Projects

"Build a 4-Channel Universal Decoder," by Fred Nichols, (Electro-Voice) *Popular Electronics*, Dec. 1972, pp. 28-32. Construction of a 4-channel decoder using the Electro-Voice EVX-44 Decoder, a very sophisticated IC—and expensive.

"Build a Crystal-Controlled Musical Instrument Tuner," by Hank Olson, *Popular Electronics*, Dec. 1972, pp. 55-57. A discrete component crystal oscillator at 901 KHz, which when divided by four IC by 2043 (2") provides 439.94 Hz and again by 2 provides 219.97, very close to 440 Hz (A above middle C), and a frequency (220Hz) suitable for tuning a bassoon.

"Sound-Activated Photoflash Attachment," by Vic Leshkowitz, *Popular Electronics*, Dec. 1972, pp. 74-75. A simple one-IC device for triggering a photo-flash in response to sound. Freezes motion.

"Vibra-Tone," by Darrell Thorpe, *Elementary Electronics*, July-August 1972, pp. 59-61, 95. Three ICs are combined with eleven contact pads to provide tones and vibrato. Block diagram and parts list.

"Put the Time on Your TV Screen," *Radio-Electronics*, Sept. 1974, pp. 33-35, 40-42, 94. Construction details, schematic, printed circuit layout, parts list and description.

"Minicomputer—Build for Under $400," *Popular Electronics*, Jan. 1975, pp. 33-38. Circuit diagrams, construction details and parts list. (Operation described in Feb. issue.)

"Build a Direct-Reading Logic Probe," *Popular Electronics*, Sept. 1975, pp. 54-56. Circuit, printed circuit board guide, parts list and instructions for building.

"Build 3 Unique Clocks," by Charles Caringella and Michael Robbins, *Radio-Electronics*, Jan. 1975, pp. 43-46 & 85, and Feb. 1975. Circuits parts lists and construction aids.

"Build a Lady's LED Time/Data Wristwatch—Only $75," by Bill Green, *Popular Electronics*, March 1975, pp. 36-41. Circuit, printed circuit board layout, parts list, construction aid.

"Methods of Matrixing for 4-Channel Sound," by Leonard Feldman, *Popular Electronics*, Jan. 1973, pp. 26-31. A description of the most important matrix systems developed through 1972.

"Introduction to Electronic Music," by Don Lancaster, *Popular Electronics*, Oct. 1973, pp. 35-37. Mostly a list of sources of information and equipment in the electronic music field.

"Audio-Visual CMOS Toy," by Joseph G. Gaskill, *Popular Electronics*, Oct. 1973, pp. 38-39. Three ICs are combined to provide switch-controlled dual tones and flashing lights LEDs.

"IC Photo Development Timer," by Robert Marchant, *Popular Electronics*, Oct. 1973, pp. 70-71. Constructional details of a photo timer using two ICs.

"Now—Build the 'Senior Scientist' Calculator," by Martin Meyer. *Popular Electronics*, Oct. 1975, pp. 33-38. Describes the construction of an advanced electronic calculator.

"Temp Cube—a Digital Indoor/Outdoor Thermometer You Build," by William J. Hawkins, *Popular Science*, Jan. 1973, pp. 95-98. Includes schematic, wiring diagram, construction details and parts list.

"Pocket Data Terminal," by Charles Edwards. *Radio-Electronics*, Jan. 1976, pp. 29-32. Block diagram, printed circuit board layout and circuit details.

"Baudot to ASCII (Converter)," by Roger L. Smith. *Radio-Electronics*, April 1976, pp. 57-59. To enable one to use Baudot or other teletypewriter as an input device for the T.V. typewriter.

"A Digital Stopclock for Short and Long Event Timing," by Michael S. Robbins. *Popular Electronics*, Jan. 1976, pp. 48-52. Circuit diagram, printed circuit board layout, parts list and operational details.

"Build a Direct-Reading Logic Probe," by R. M. Stitt, *Popular Electronics*, Sept. 1975, pp. 54-56. Circuit diagram, parts list, pc board layout and construction details.

Glossary of Terms

A/D CONVERTER—analog to digital converter, converts analog voltages to their digital equivalents.

ACTIVE FILTER—R-C circuits and amplifiers combined to simulate an L-C filter.

ACTIVE RESONATOR—the R-C and amplifier circuits simulating an L-C tuned circuit (part of an active filter).

ADDER—a circuit producing an output which is the sum of its inputs.

ANALOG IC—same as linear IC. (Provides an output which is a linear function of its input.)

AND GATE—a gate which produces an output (1) when and only when all its inputs are high: (1) $A \cdot B = C$.

ASTABLE MULTIVIBRATOR—a multivibrator (flip-flop) which oscillates between two states.

BINARY—the system of counting by radix or base 2.

BIPOLAR—transistor or its equivalent in an IC which conducts by means of both holes and elections (transistors, *not* field effect).

BIT—a single logic state 1 or 0.

BITE—a number of logic states, typically five.

BREADBOARD—a board or board-like panel for assembly and testing of circuits on an experimental or temporary basis.

BUFFER—means for storing temporarily or otherwise adapting a circuit or device to another otherwise incompatible circuit or device.

CHARGE COUPLED (CCD)—semiconductor cells electrostatically coupled and characterized by the storage of patterns of pulses.

CHOPPER STABILIZED AMPLIFIER—an amplifier in which the dc input is chopped simulating ac in order to overcome amplifier dc drift.

CIRCUIT DENSITY—the number of equivalent transistors per unit area of a IC chip.

214

CLOCK—a series of equally spaced pulses used for timing the operation of circuits which require synchronous operation.

CLOSED LOOP GAIN (LOOP GAIN)—the actual gain of an operational amplifier circuit when a feed-back circuit is connected.

COMMON MODE-REJECTION RATIO (CMRR)—the effective attenuation at an operational amplifier output of common mode inputs.

COMPLEMENTARY INSULATED-GATE FIELD-EFFECT TRANSISTOR LOGIC (COS/MOS OR C/MOS)—logic employing complementary transistors of the insulated gate field effect type.

COS/MOS—complementary metallic oxide insulated gate field effect transistors or IC.

C/MOS—Same as COS/MOS.

COUNTER—a series of cascaded flip-flops changing states progressively in response to repeated input counts.

CHANGE STATE—to go from logic state 1 to logic state 0 or the inverse.

COMMON MODE—voltages which equally affect both inputs of an operation amplifier.

CHIP—a thin slice of semiconductor material on which ICs are formed.

CROSS-OVER DISTORTION—in a push-pull system the non-linearity resulting from imperfect transition from positive half cycles to negative half cycles.

D/A CONVERTER—digital to analog converter; converts digital numbers to their analog equivalents.

DATA AMPLIFIER—a high performance amplifier meeting predetermined performance standards for data processing.

DATA SELECTOR—a plurality of gates connected between a source of data and utilization means: for example, two control lines, the condition of which determine the gates to be enabled to pass data.

DIE—a chip cut into small rectangular pieces, each carrying an individual IC.

DIFFERENTIAL COMPARATOR—a differential circuit for indicating when two input signals are essentially equal, as in a differential pair.

DIFFERENTIAL PAIR—a pair of transistors sharing a common emitter circuit but with two independent base inputs.

DIODE-RESISTOR LOGIC (DRL)—logic gates formed of combinations of diodes and resistors, such as OR and AND gates.

DIODE GATE—two or more diodes arranged to conduct desired signals and to block undesired signals.

DIRECT-COUPLED UNIPOLAR LOGIC (DC UTL)—direct-coupled logic employing field effect transistors.

DISCRETE COMPONENT—a single component such as a transistor or resistor not integrated with other components.

DOPED MATERIAL—semiconductor material to which a predetermined amount of impurity has been added.

DRAIN—the high potential or load end of a field effect transistor or IC such as the collector of a transistor.

D TYPE FLIP-FLOP—a flip-flop for accepting and holding data.

DUTY CYCLE—the percentage of time a signal or device is turned on or operating at full output.

DRIFT—the change of a circuit parameter with time.

DEBUGGING—to locate the cause of faulty operation and to provide a cure.

DIGITIZER—means for converting analog signals to digitial signals according to a predetermined law of conversion, permitting decoding to reconvert to the original analog form.

DECODER—means for converting a pattern of pulses as used in a transmission or processing circuit or device to a new pattern.

EXCLUSIVE OR—a gate providing high state output (1) if any one input is high (1) but not if more than one input is high (1); expressed by the equation $A \cdot \overline{B} + \overline{A} \cdot B = C$.

DIGITAL INTEGRATED CIRCUITS—an integrated circuit intended to be operated on (1) or off (0), i.e., in one of two possible states of conduction.

DIGITAL MODE—ICs operating in one of two states, off or on (0 or 1).

DYNAMIC OR VOLATILE CELL (MEMORY)—a cell which holds a given state temporarily and must be periodically refreshed.

De MORGAN'S LAWS—the equivalents expressed by the two equations: (1), $\overline{A} + \overline{B} = A \cdot B$ and (2), $\overline{A} \cdot \overline{B} = A + B$.

DEPLETION MODE—a field effect transistor or IC which passes maximum current at zero gate potential and a decreasing current with applied gate potential.

EMITTER-COUPLED LOGIC (ECL/MECL)—transistor logic in which multiple inputs are connected to multiple emitters of an input transistor.

ENCODER—means for changing a pattern of pulses from an initial pattern to a new pattern adapted for use or processing in a following circuit or device.

ENHANCEMENT MODE—a field effect transistor or IC which passes little or no current at zero gate potential and an increasing current with an increase in gate potential.

EPITAXIAL—a semi-conductor grown from a single-crystal by vapor phase growing with controlled resistivity and thickness.

EMITTER FOLLOWER—the transistor amplifier circuit in which input is applied to the base and output taken from the emitter.

FAN-IN—the provision for a number of input lines to a given device over those normally provided.

FAN-OUT—circuits where a given IC driver or is connected to the inputs of more than one further IC device.

FIELD-EFFECT TRANSISTOR LOGIC (JFET)—logic employing field effect transistor.

FLIP-FLOP—cross-coupled transistors or gates providing two states of conduction. Further characterized as astable, bistable or monostable.

FREQUENCY SHIFT KEYING—digital information is converted to two discrete frequencies, one representing 0 and the other 1.

FULL ADDER—an adder which accepts two inputs as in a half adder and produces an output, but in addition accepts a carry input (a third input).

GATE EXPANDER—means for adding inputs to a gate. called an expandable gate.

GROWING—semiconductor formation from a melt; a seed crystal starts growing and resulting ingot is slowly withdrawn, forming long solid crystal.

GYRATOR—a device providing 180-phase shift in one direction relative to the other direction of signal passage.

GATE—the control element of a field effect transistor or IC as the base of a transistor.

HALF ADDER—an adder which accepts only two inputs and produces an output but cannot handle a carry input.

HARDWARE—the actual components which go to make up a given system.

HEAT-SINK—heat radiating or dissipating device for cooling transistors or ICs.

HIGH SPEED PRINTER—a signal responsive alpha/numeric printer capable of printing computer output signals at rates of the order of 300 characters per second or greater.

HIGH-VOLTAGE LOGIC (HTL)—a class of transistor logic having an offset of about 6 volts in its input, intended to increase noise immunity.

HYBRID AMPLIFIER—an amplifier employing both discrete and integrated circuits.

IMPURITY (INDIUM OR GALLIUM)—materials added to a semiconductor melt for producing either N or P material.

INSTRUMENTATION AMPLIFIER—a precision amplifier meeting predetermined high standards of its parameters.

INSULATED GATE FIELD EFFECT TRANSISTOR LOGIC (MOS/FET)—logic employing insulated gate field effect transistors.

ISOLATION—separated as by insulating means preventing disturbance by electrical signals.

ION IMPLANATION—increasing conductivity of a semiconductor by high speed ion bombardment.

INTEGRATED CIRCUITS—a plurality of semiconductor elements grown, etched or deposited on a substrate and packaged as a multi-element active device.

INVERTER—IC whose function is simply to invert, i.e., change 0 to 1 or 1 to 0.

INVERTING INPUT—in an operational amplifier the input terminal at which an input signal produces an inverted phase output. Degenerative feedback goes from the output to this terminal.

INHIBIT—to prevent operation of a gate, for example, by means of a blocking signal.

INPUT VOLTAGE OFFSET—in an operational amplifier, the effective non-zero input voltage due to unbalance in input transistors.

INPUT CURRENT OFFSET—in an operational amplifier, the effective non-zero equivalent input current due to unbalanced input current drawn by the input circuit transistors.

INPUT IMPEDANCE—the input impedance of an amplifier.

INTERMODULATION DISTORTION—the 3rd and higher order odd power distortion which causes two signals to cross-modulate.

JFET—junction field effect transistor or IC; the gate is uninsulated from the channel and draws current.

JK FLIP-FLOP—a flip-flop meeting the equation $Q+ = \bar{Q}J + Q\bar{K}$; counts clock input pulses when $JK = 1$.

LATCH—a monostable flip-flop which when pulsed assumes and holds a predetermined state.

LIGHT-EMITTING DIODE (LED)—a semiconductor diode which emits light when excited.

LINEAR MODE—ICs operated in portions of their characteristics where the output is a linear function of the input.

LINEAR INTEGRATED CIRCUITS—an integrated circuit intended to be operated over a range in which the output is directly proportional to the input.

LOCKING RANGE—the frequency range over which a phase-locked loop oscillator will follow an input signal in locked frequency.

LOGIC 0—the low or zero state of an input or output where logic 1 is a positive voltage (positive logic) or the negative maximum input or output where logic 1 is near zero (negative logic).

LOGIC 1—the maximum positive voltage state where logic is near zero or the near zero state where logic 0 is maximum negative. (See above.)

LOGIC STATE—the high (1) or low (0) state of a circuit.

LARGE SCALE INTEGRATION (LSI)—integration such as computer chips and large memories composed of more than 100 gates or the equivalent.

MASKING—microscopically small cut-outs made by optical reduction of large drawings and projected on a semiconductor substrate for defining etching patterns of IC component parts.

MEDIUM SCALE INTEGRATION (MSI)—integration such as registers, counters and small memories composed from 20 to 100 gates or so.

MEMORY—means for holding digital information on a temporary basis or permanent basis (see RAM and ROM).

MIL SPECIFICATIONS—military standards specifications.

MONOSTABLE MULTIVIBRATOR—a multivibrator (flip-flop) stable in only one condition of conduction.

MULTIPLEX—the transmission of more than one series of signals over a single channel.

MUSIC SYNTHESIZER—ICs capable of programming musical tones in a predetermined or random fashion.

MOS/FET—metallic oxide insulated gate field effect transistor or IC.

NAND GATE—a gate which produces a low output (0) when and only when all inputs are high (1) $A \cdot B = C$.

N-CHANNEL—field effect transistor or IC in which the channel between source and drain is N doped semiconductor, generally silicon.

NOISE—non-signal disturbances generated by thermal agitation or from other sources in an IC circuit.

NOISE IMMUNITY—the difference between actual signal levels in a circuit and the levels required to produce response in the gate or other device.

NON-INVERTING INPUT—in an operational amplifier, the input terminal at which an input signal produces an in-phase output.

NON-LINEAR DISTORTION—the departure of the output of an amplifier from strict linear function of its input, generally expressed in percent distortion.

NOR GATE—a gate which produces a low output (0) when either or both inputs are high (1) $A + B = C$.

NEGATIVE LOGIC—inverted positive logic with the 1 state more negative than the 0 state.

OPERATIONAL AMPLIFIER (OP AMP)—a high-gain stable amplifier designed to operate with a substantial amount of degenerative feedback which in turn determines the actual gain.

OPTO-COUPLER (OPTICAL COUPLER)—a light source such as an LED enclosed with a light-sensitive semiconductor diode or transistor, a circuit-isolating device.

OR GATE—a gate which produces an output (1) when either or both inputs are high (1) $A + B = C$.

OPEN LOOP GAIN (LOOP GAIN)—the actual gain of an operational amplifier with no feedback.

OUTPUT VOLTAGE SWING—the range of output voltages from plus to minus of which an amplifier is capable at rated bias voltage.

PACKAGE DISSIPATION—the power, generally in milliwatts or microwatts, consumed by a given IC during normal operation. It depends on the states of the active elements at a given time.

P-CHANNEL—field effect transistor or IC in which the channel between source and drain is P doped semiconductor, generally silicon.

PAD—metal deposits on the IC chip providing means for connecting leads going to package pins.

PHASE-LOCKED LOOP (PLL)—a voltage-controlled oscillator (VCO) and a comparator feeding back a control signal forcing the oscillator to lock onto an input signal.

POSITIVE LOGIC—logic in which the 1 state is more positive than the 0 state.

PROGRAMMABLE READ ONLY MEMORY (PROM)—a read only memory: the states of the cells being alterable in a manner which depends on the cell structure.

PROPAGATION DELAY—the time it takes for the output of a device to respond to an input signal.

Q—symbol for a transistor.

RANDOM ACCESS MEMORY (RAM)—memory composed of cell which can be accessed to write or read data.

READ ONLY MEMORY (ROM)—memory composed of cells the state of which are predetermined and can be read but not altered.

REGISTER—a series of cascaded flip-flops capable of receiving pulses in series or parallel and retaining same until cleared or reset.

RESISTOR-TRANSISTOR LOGIC (RTL)—logic gates formed of resistors and transistors.

RESISTOR-CAPACITOR-TRANSISTOR LOGIC (RCTL)—logic gates formed of combinations of resistors and transistors in which capacitors are placed across the resistors to speed up response.

RING COUNTER—a series of flip-flops with the last returned to the first so that count states circulate.

RIPPLE COUNTER—a counter comprising cascaded flip-flops. Used as a divider.

R-S FLIP-FLOP—a flip-flop which meets the equation $Q+ = S + Q\bar{R}$ with the restriction $RS=0$.

SAMPLE-AND-HOLD—a circuit which stores a sampled voltage for a short time, generally in the form of a charge on a capacitor.

SCHMITT TRIGGER—two transistors regeneratively cross-coupled so that the circuit flips ON at one input level and OFF at a slightly lower input level.

SEED CRYSTAL—a crystal used to start the growing of a large semiconductor ingot.

SHIFT REGISTER—a register in which input pulses cause the register content to step sequentially.

SIGNAL RECONDITIONING—the act of partially or completely restoring the original form of a distorted signal.

SLEWING RATE—the volts-per-unit time which an amplifier output changes in response to a step input voltage change.

SILICON-ON-SAPPHIRE (SOS)—ICs formed by depositing or growing silicon transistor elements on a sapphire substrate.

3-STATE LOGIC—logic which can assume the two usual 0 or 1 states and in addition a third or OFF-LINE state.

STATIC CELL (MEMORY)—a memory cell which holds a given state without being refreshed.

STORAGE CELL—an IC capable of storing one bit.

STORAGE—means for storing large amounts of digital information such as a magnetic tape or magnetic disk storage.

SENSE AMPLIFIER—an amplifier used for sensing the OFF or ON state of magnetic core memory cells.

STROBE—an enabling pulse usually applied to a gate for releasing or reading out information being held.

SUMMING JUNCTION—the input terminal of an op amp which is inverting and has both input and feedback connected to it.

SWITCHING TIME—the time taken for a gate output to respond to a signal applied to its input.

SPEED—the rate at which signals are propagated or states are changed in response to input signals.

SOURCE—the low potential end of a field effect transistor corresponding to the emitter of a bipolar transistor.

SMALL SCALE INTEGRATION (SSI)—integration such as gates or flip-flops composed of the equivalent of up to about 20 transistors or gates.

SUBSTRATE—the base chip material on which ICs are grown, deposited or otherwise integrated.

TOGGLE—a flip-flop the output of which changes state for each input pulse (clock pulse).

TRANSIENT RESPONSE—the response of a circuit to a step input pulse resulting in over-shoot or ringing.

TRUTH TABLE—a table showing all possible combinations of input levels to a gate or other circuit and the resulting output levels.

UNIPOLAR—transistor or its IC equivalent which conducts by means of holes or electrons, not both (field effect device).

VOLTAGE-CONTROLLED OSCILLATOR (VCO)—an R-C oscillator in which the frequency is controlled by means of an applied voltage.

WHITE NOISE—a continuous spectrum of random noise.

WIDEBAND AMPLIFIER—amplifiers having a wide frequency response of substantially uniform gain, generally without feedback.

WORD—a group of coded logic states, generally 8 to 16, of a predetermined number forming a ''word'' in a predetermined code.

Bibliography

Electronics for 1971
Electronics
McGraw-Hill, Inc.
1221 Avenue of the Americas
New York, N.Y. 10020

Issue	*Pages*	*Subject*
April 26	38-47	H-P DVM
April 26	48-52	Case for ECL
Nov. 8	91-94	Testing ICs
Nov. 8	61-76	Opto-electronics
July 5	53-56	Charge storage
July 5	62-66	MOS/LSI
July 5	70-72	Level shift
July 5	97	Handling ICs
Sept. 27	91	Automatic testing
Sept. 27	92	LSI tester
Oct. 25	82-88	Memory delay specs.
Nov. 8	85-89	MOS shift register
Nov. 8	91-94	Test specs.
Nov. 8	121	Breadboarding
Aug. 4	171	Op amp testing
April 12	92-95	Laser beam fault-finder
April 26	48-52	Case for ECL
April 26	38-47	ICs in instruments
July 19	76-79	Denser bipolar ICs
Aug. 2	46-49	Hall-effect devices
Aug. 2	54-57	Language for test engineers
Aug. 2	58	LED null indicator
Aug. 2	69-75	Single-cell MOS memory
Sept. 13	125-126	Multiplier circuit

Electronics for 1971 (continued)

Issue	Pages	Subject
Nov. 22	56-61	2 × 4 bit multiplier
Nov. 22	62-66	FM stereo decoder
Nov. 22	83-84	Needed board tester
Dec. 6	70-75	Discussion of reliability
Dec. 6	86-91	Charge-coupled
Dec. 6	92-94	Transient protection
Dec. 20	44-48	MOS processing
Dec. 20	49-53	PCM data acquisition
Dec. 20	80	Test generator
April 12	107-108	Shipboard EDP
April 12	68-73	Molybdenum MOS
April 12	75-86	Packaging methods
Aripl 12	92-95	Laser testing ICs
April 12	97-100	IC watches
July 19	60-65	Varactor chip
July 19	71-72	ECL to TTL interface
July 19	72	Tunable active filter

Electronics for 1972

Issue	Pages	Subject
Aug. 28	63-77	State-of-the-art RAMs
Aug. 28	94-95	New ECL
Jan. 3	103	Data acquisition
Jan. 31	66-69	Digital clock
Jan. 31	89-90	MSI multiplexer
April 10	69-70	Beam leads
April 10	77-78	Silicon on sapphire
April 10	85-91	JFETS replace VTs
April 10	93-97	C/MOS watch
April 10	102-108	Functional trimming
April 10	121-123	Chip processor
April 10	124-126	Multipurpose tester
June 5	83-98	Bridging digital to analog ICs
June 5	103-107	Three technologies on a chip
June 5	120-121	Time delay
June 12	111-113	LSI-ROM
Sept. 11	109	Gray-code generator

Electronics for 1972 (continued)

Issue	*Pages*	*Subject*
Sept. 25	108-109	Special ring counter
Sept. 25	113-116	Silicon on sapphire
Sept. 25	117-120	16 bit D/A conversion
April 24	85-90	Ion implant MOS
April 24	106-110	Active resonators
Aug. 14	93-107	Consumer IC uses
Aug. 14	116-119	Testing complex logic circuits
Aug. 14	124-126	Temperature compensated oscillator
Sept. 11	102-105	LSI A/D converter
Sept. 11	112-119	Aerospace computer
Sept. 11	121-126	ROM for fuel injection
Sept. 25	121-122	Phase-locked loop
July 3	81-85	Conversational software (AEDCAP)
July 3	90-95	Weather telemetry
July 3	96-104	Collector diffusion isolation
Oct. 9	96-99	Caution re. 16-bit A/D converters
Oct. 9	104-109	Complex circuit testing
Oct. 9	112-117	Fast ECL comparator
Oct. 9	120-121	Low drift ICs
Aug. 14	115	Automated testing
Aug. 14	120-123	Complexlogic test patterns
Jan. 3	89-92	Self-aligning MOS process

Electronics for 1973

Issue	*Pages*	*Subject*
Feb. 15	112-114	ROM
Feb. 15	115-117	MOS Memory
March 29	94-95	Breadboard
June 7	89-104	Thick and Thin Films
June 7	108	Sense Amplifier
June 7	117-120	C/MOS Applications
June 7	123	Two Chip Chopper
June 21	128-132	555 Timer
Aug. 2	101-104	Sample-and-Hold
Aug. 2	75-90	Special Report
Aug. 2	91-94	Hybrid Power Supply

Electronics for 1973 (continued)

Issue	Pages	Subject
Oct. 11	119-121	Ceramic DIP
Oct. 11	129-130	LSI Tester
Oct. 25	102-104	Communications
Nov. 8	89-93	Logic Clip

Electronics for 1974

Issue	Pages	Subject
Jan. 10	100-101	n-MOS
Jan. 24	81-86	Blind Aid
Feb. 7	63	Auto Monitor
March 21	111-116	Microprocessor
April 4	125-130	D/A Converter
May 2	33-34	MOS Computer
May 2	100-101	Bipolar RAM
May 2	115-116	Access Time
May 2	129-130	4,096 bit MOS RAM
May 16	123-128	Active Filters
May 30	120-121	In-Circuit Test
Aug. 8	91-101	Charge Coupled
Aug. 22	87-94	Slow Taped Speech
Aug. 22	98-99	IC Tester
Oct. 3	111-118	I^2L Process
Dec. 12	96-104	Solid State Watches
Dec. 26	87-93	Microprocessor

Electronics for 1975

Issue	Pages	Subject
Jan. 9	103-107	C/MOS Specs.
Jan. 23	90-95	Microprocessor Part 1
Feb. 6		Microprocessor Part 2
March 6	87-92	Microprocessor
March 20	101-106	Microprocessor
May 1	75-81	Digital System Testing
May 15	77-88	C/MOS (Special Report)

Electronics for 1975 (continued)

Issue	Pages	Subject
June 12	89-98	Word Processing
June 12	115-120	Sapphire MOS
June 26	105-109	Lambda Diode
July 10	81-92	Bipolar LSI
July 10	107-111	Timing LSI-MOS
July 24	102-106	C/MOS RAMs
Aug. 21	74-81	ICs in Cameras
Sept. 4	89-95	Bipolar LSI
Sept. 4	98-102	IN-Circuit Testing
Sept. 4	110-114	IC Organ

IEEE Spectrum

The Institute of Electrical and Electronics
Engineers, Inc.
345 East 47th Street
New York, N.Y. 10017

Issue	Pages	Subject
July 1972	63-66	A/D & D/A Converters
Sept. 1972	47-56	A/D & D/A Converters
June 1973	22-28	Cameras
June 1973	30-35	Industrial Noise Immunity
Sept. 1973	83-92	Error Detection
Oct. 1973	40-47	Automatic Testing (bib)
Oct. 1973	40-47	Diagnostics for Logic Networks
April 1974	34-41	Pocket Calculators
Sept. 1974	44-52	Automatic Testing
Nov. 1974	28-35	Memories
Nov. 1974		Instrumentation Issue
Dec. 1974	63-70	Logic Analyzers
Feb. 1975	40-41	Memories
Aug. 1975	40-45	LEDs

Scientific American

Scientific American, Inc.
415 Madison Avenue
New York, N.Y. 10017

Issue	Pages	Subject
April 1973	63-69	Ion Implantation
Feb. 1974	22-32	Charge-Coupled Devices
April 1974	28-35	Integrated Optics

Computer Design

Computer Design Publishing Corporation
221 Baker Avenue
Concord, Mass. 01742

Issue	Pages	Subject
May 1973	99-106	C/MOS Logic
Jan. 1974	65-77	Selection of Memories
Feb. 1974	42-52	Automotive Standards

Electronic Products

United Technical Publications, Inc.
654 Stewart Avenue
Garden City, N.Y. 11530

Issue	Pages	Subject
July 15, 1974	39-49	Logic Analyzers

Index

A

Actual circuits of ICs (*see* Circuits [internal] of ICs)
Adders:
  basic to logic computers, 90
  full-adder, 90-91
  full-adder truth table, 91
  half-adder, 90
  half-adder truth table, 90
Analog IC, 21
Analog-to-digital (A/D) converters:
  bipolar, 95
  improving accuracy, 95-96
  ramp comparison, 93
  real world interface to ICs, 92
  signal conditioning, 95-96
  staircase ramp, 94
  unipolar, 95
AND gate (*see* Gates, AND)
Application notes for ICs, 153-155
Automatic test equipment (ATE):
  software assistance, 186
  user recommendations, 186-187
Automatic test systems (ATS):
  hardware realities, 185-186
  ideal system, 185
  user recommendations, 186-187
Availability of ICs, 150

B

BCD definition, 88
BCD-to-decimal decoder, 89-90
Bibliography, hobby projects, special, 212-213
Bibliography, main:
  Computer Design, 229
  Electronics for 1971, 224-225
  Electronics for 1972, 225-226
  Electronics for 1973, 226-227
  Electronics for 1974, 227
  Electronics for 1975, 227-228
  Electronic Products, 229
  IEEE Spectrum, 228
  Scientific American, 229

Block diagrams convert to hardware, 115
Boolean algebra:
  equations, 28-29
Breadboarding:
  advanced form, 106-107
  auxiliary devices, 108
  check and recheck, 109
  critical areas, 113
  defining, 102
  exclusive-OR experiment, 104-106
  experimenter's kits, 108
  final testing, 113
  555 timer in, 110-112
  for DIPICs, 103, 106-107
  heat-sinking, 113
  high frequencies, at, 112
  J-K flip-flop, 104-105
  laboratory, 106-107
  LED indicators, 103-105
  logic state indicators, 109
  noise problem, 109
  over-current in, 110
  problems, 109
  reasons for, 102
  regulated power supply, 110
  soldering, 109
  temporary circuit, 102
  test points, 112
  timer for, 108, 110-112
  troubles, 109-110
  types, 103-104
  why, 102
  with two DIP sockets, 104-106
Buffer/inverter (RTL), 45

C

Calculators, pocket (*see* Pocket calculators)
Characteristics of ICs vary, 153
Charting to simplify:
  chart form of truth table, 130
  charting defined, 130
  charting illustrated, 130-135
  directly from chart to hardware, 134-135
  double inversion, 134
  more than three variables, 133-134

233

Charting to simplify: *(cont.)*
  resulting gate combinations, 132-135
  rules for charting, 131-134
  simple chart, 130-131
  simplifying equations, 132-135
  single inversion, 135
  three variables, 131-133
  two variables, 130-133
Choosing IC family, 152
Circuit density, 22
Circuits (internal) of ICs:
  basic TTL NAND gate, 36
  C/MOS NAND gate, 56
  C/MOS NOR gate, 57
  C/MOS three-state gate, 58-59
  C/MOS transmission gate, 58
  diode AND gate, 33
  discussion of, 21-22
  DTL NOR gate, 47
  ECL basic gate, 50
  ECL 5-input gate expander, 52
  ECL 3-input expandable gate, 51
  flip-flop basic circuit, 61
  HTL NAND gate, 54
  MOST/FET NAND gate, 55
  negative logic OR and AND gates, 38
  RTL basic gate (NOR), 44
  RTL buffer/inverter, 45
  transistor AND gate, 35
  transistor NOT gate, 29, 35
  transistor OR gate, 34
  TTL expandable AND-OR-INVERT
    gate, 49
  TTL gate expander, 49
  TTL NAND gate, 48
Clock:
  defined, 71
  frequency, 71-73
  pulse phase, 73
  synchronizing local clock, 72
Clocked systems, 72-73
Clocks (*see* Electronic clocks)
C/MOS:
  application notes, 154-155
  defined, 56
  gates, 56-59
  interfacing C/MOS, 117-121
  noise margin, 115-116
Code converters, 87-90
Compatibility of logic types, 114

Converters:
  accuracy of D/A, 100-101
  analog-to-digital, 92-96
  analog-to-digital (A/D) defined, 92
  back to real world (D/A), 97
  binary-weighted resistors, using, 97-99
  bipolar D/A, 99-100
  D/A defined, 97
  D/A easier than A/D, 97
  D/A ladder network advantage, 98
  D/A practical problems, 97-98
  digital-to-analog, 97-101
  improving accuracy, 95-96
  interface real world, 92
  ladder network, 98
  ladder network improvement, 98-99
  most common D/A method, 97
  most significant bit, 98
  operational amplifiers, in, 97-101
  practical applications of D/A, 101
  practical questions about, 92
  ramp comparison, 93-94
  resistor ratios in, 101
  resistor temperature coefficient, 101
  resolution, 96
  sample and hold, 96
  signal conditioning in, 95-96
  specifying A/D, 96
  staircase ramp, 94-95
  12-bit accuracy in D/A, 100
Cooling ICs costly, 151
Cost factor in TCs, 150-151
Counters:
  as interval timers, 78
  combine gates and flip-flops, 74-78
  divide by any factor, 74-78
Critical applications of ICs, 153

D

Data selector, 90
d-c testing, 167-170
Decoding and encoding:
  binary (BCD) to decimal, 89-90
  binary to gray code, 87-88
  gray code to binary, 87-88
Digital ICs in hobby projects (*see* Hobby
    projects)
Digital IC testing (*see* Testing ICs, and
    Specification, testing to)

DIP ICs, breadboarding, 103, 106-107
DIP sockets, 104-106
Dividers:
  count forcing, in, 75
  factors, 60, 74-78
  gates and flip-flops, 75-77
  using counters, 74-78
  using flip-flops, 74
Double inversion charting, 134-135
DTL noise margin, 115-116

E

Electronic clocks:
  block diagrams, 141, 144
  crystal oscillator, 141
  display, 141, 142, 144
  50/60 cycle source, 141, 144
  liquid crystal display, 142
  MSI and LSI combined, 143
  operation, 141-144
  readout, 141-142, 144
  readout multiplexing, 141, 142, 144
Electronic watches:
  block diagram, 145
  crystal frequency standard, 143
  display multiplexing, 143
  liquid crystal display, 145
Encoding and decoding:
  binary (BCD) to decimal, 89-90
  binary to gray code, 87-88
  gray code to binary, 87-88
Equations (see Charting to simplify)
Equivalent circuits aid to understanding ICs,
  42-43
  (also see Circuits [internal] of ICs)
Exclusive-OR, breadboarding, 104-106
Exclusive OR gate, 39-40
Expanders, gate:
  C/MOS fan-out, 124
  design factors, 122
  DTL fan-out, 123
  ECL fan-out, 123
  ECL gate, 125
  fan-in gate, 124-126
  fan-in requirements, 121-124
  fan-out requirements, 121-124
  RTL fan-out, 122-123
  RTL gate, 124-125
  TTL fan-out, 123-124

Expanders, gate: (cont.)
  TTL gate, 49, 125-126
Experimenter's kits, 108
555 Timer, circuit and applications, 110-112

F

Flip-flops:
  basic circuit, 61
  basic R-S flip-flop, 62
  clocking, 71-73
  definition, 61-62
  J-K, 66-67, 104-105
  propagation delay, 69
  pulsed, 70
  random access memory, as, 79-84
  recovery delay, 69
  reset, 63
  ring counter, connected as, 79
  R-S-T (toggle), 66
  R-S truth table, 63
  set, 63
  shift registers, form, 78-79
  simple, 63
  synchronous, 64-65
  synchronous operation, 68-70
  timing diagrams, 65
  toggle, 66
  transmission delay, 70
  type D, 65
  uses, 68

G

Gates:
  basic building blocks, 43
  C/MOS NAND gate, 56-57
  C/MOS NOR gate, 57
  complex logic, form, 60
  diode AND, 33
  diode OR, 33
  expanders, 49, 51, 52
  inverting, 36-37
  negative logic, 51, 55
  paralleling gates, 59-60
  permutations of, 41
  positive NAND, 36-37, 47-48, 50, 54-56
  positive NOR, 37, 47, 50, 57
  three state, 58-59

Gates: *(cont.)*
  transistor AND, 35
  transistor NOT, 35
  transistor OR, 34
  wired AND/OR, 59-60
Glossary of Terms, 214-223
Ground connections in logic diagrams, 57
Ground loops broken by opto-coupling, 126

H

Heating at high frequencies in ICs, 151
Heating in C/MOS ICs, 151
Hermetically sealed ICs, 152
History of ICs:
  before the IC, 19
  composed of transistors, 21
  field effect ICs, 24
  finest IC, 19-20
  growing ICs, 22
  new technology, 22
  packaging, 27
  SSI/MSI/LSI defined, 26
  switch/gate function, 21
Hobby projects:
  (see Sound level indicator, and Voice
    controlled device)
  analog projects, differ from, 189
  applying power, before, 192-193
  availability of parts for, 190-191
  before starting, 189-190
  bibliography, special, 212-213
  digital clocks a "natural," 189
  equipment needed, 190-192
  ICs great time savers in, 191
  inventing a new, 193-204
  inventing start to finish, 193-204
  kits, 188, 190, 192
  learn by doing, 192
  music synthesizer block diagrams,
    207-209
  music synthesizer kits, 210
  music synthesizers, 206-210
  potential and interest, 188-189
  proper place to build, 192
  questions before starting, 189
  tips on building, 189-193
  tools required, 191
  unused gates, 193
  warning, 211
HTL (high-threshold-logic), 119

I

In-circuit testing, 172-175
Interfacing ICs:
  C/MOS driving DTL, 118-119
  C/MOS driving TTL, 118-119
  C/MOS fan-out, 124
  C/MOS interfacing ECL, 120-121
  C/MOS interfacing HTL, 119
  driving opto-couplers, 129
  DTL driving C/MOS, 117
  DTL fan-out, 123
  ECL expanders, 125
  ECL fan-out, 123
  ECL with C/MOS, 120-121
  fan-in gate expanders, 124-126
  fan-in requirements, 121-124
  fan-out, 117
  fan-out factors, 122
  fan-out requirements, 121-124
  HTL interfacing C/MOS, 119
  important factors, 117
  input current requirements, 117
  input/output devices, 121
  isolation by opto-coupling, 127
  LED in opto-coupler, 127
  logic swing, factor, 117
  no feedback in opto-coupler, 130
  noise margins, 117
  opto-coupler configurations, 127-129
  opto-coupler construction, 126-130
  opto-coupler fast response, 127
  opto-coupling, 126-130
  opto-coupling breaks ground loops, 126
  power supply considerations, 117
  problems solved by opto-coupling,
    126-130
  pull-up resistors, 117-118
  RTL expander, 124-125
  RTL fan-out, 122-123
  translators, 121
  TTL driving C/MOS, 117
  TTL expanders, 125-126
  TTL fan-out, 123-124
Intermediate region, 115-116
Interval timers, 78
Inventing a new hobby device *(see* Voice
    controlled hobby device)
Inventing hobby projects, 193-204
Inverter, basic (RTL), 44
Isolation by opto-coupling, 127

J

J-K flip-flop, 66-67, 104-105

K

Kits for hobby projects, 188, 190, 192

L

Large scale integration (LSI):
    clocks a "natural," 140
    C/MOS dense but slow, 137
    computer programs for, 22
    defined, 26
    dynamic memory, 147
    electronic clocks, 140-144
    electronic watches, 143, 145
    IC and core memories compared, 146
    IC RAM a challenge to, 145
    IC RAM capacity expands, 147
    IC RAM compared to other methods,
        146
    IC RAM made practical, 136
    IC RAM speed, 146
    Ion implantation speeds MOS, 137
    limitations to size, 136-137
    many applications, 136
    pocket calculators, 137-140
    revolutionized IC field, 136
    16 bit memory, 146
    thousands of transistors, 22
    thousands of transistors on a chip,
        136-137
    two FET modes on one chip, 137
Light emitting diode (LED):
    LED in opto-coupler, 127
Liquid crystal displays, 142, 145
Local clock, 72
Logic:
    Boolean algebra, 28-29
    defined, 28
    DeMorgan's Laws, 41
    explains IC operation, 28
    negative, defined, 38
    positive, defined, 38
    symbols, 29-30
    transistor, 36
    truth tables, 31-32, 39-41
Logic analyzer, 176-184

Logic clip circuit, 175
Logic clips, 173-175
Logic ICs (*see also* Gates):
    basic gates, 30-31, 43
    capacitor/resistor (RCTL), 46
    complementary metal oxide (C/MOS),
        56-59
    diode-transistor (DTL), 46-47
    direct coupled unipolar (DC UTL), 54
    emitter-coupled (ECL/MECL), 50-53
    evolution, 42
    high voltage (HTL), 52-54
    metal-oxide-semiconductor (MOS/FET),
        54-55
    resistor-transistor (RTL), 43-46
    three state, 58-59
    transistor-transistor (TTL), 46-49
    types, 43
Logic probes, 172-173
Logic state indicators, 109, 103-105

M

Manufacturing ICs:
    basic process, 22
    epitaxial, 24
    field effect ICs, 24-25
    five steps, 22-23
    interconnections, 24
    packaging, 23
    silicon crystal, 22
Medium scale integration (MSI):
    defined, 26
MIL specifications for ICs, 152
Multiplexer truth table, 86-87
Multiplexers, 86-87
Multiplexing circuit, 86
Music synthesizers, 206-210

N

Noise immunity:
    C/MOS, 115-116
    DTL/TTL, 115-116
    specifications, 115-116
Noise margin (diagrams), 115-116
Non-critical IC applications, 153
Opto-coupler (*see* Interfacing ICs)

P

Packaging ICs:
 dissipation of IC packages, 152
 dual-in-line (DIP), 27
 early methods, 27
 flat-pack, 27
 sockets, 27
Performance limits for ICs, 150
Photographs:
 breadboard, 103
 breadboard, laboratory, 107
 circuit box (two DIP), 104
 circuit box, OR-gate set-up, 106
 circuit box, push button, 105
 CRO of BCD counter state flow, 181, 183
 CRO of transient, 182
 head of common pin/IC, 25
 logic state analyzer, 180
 microprocessor chip, 23
 plan view of early IC, 20
Plactic package ICs, 151-152
Pocket calculators:
 block diagram, 137-138
 chip functions, 137-139
 display functioning, 138-139
 entering input, 139
 functions, 139
 input register, 137-139
 keyboard, 139-140
 memory, 140
 overflow, 138
 pre-recorded program, 140
 single IC, 137
 what goes into, 137-140
 wide use of LSI, 137
Power dissipation in ICs, 151
Power requirements for ICs, 151
Power sources for ICs, 151
Programmable read only memory (PROM):
 alteration methods, 85-86
 defined, 85
 P/ROM, called, 85
Propagation delay, 68-70

R

Random access memory (RAM):
 defined, 79
 IC capability expands, 147

Random access memory (RAM): *(cont.)*
 IC LSI, 145-146
 speed compared, 146
Read only memory (ROM):
 defined, 84
 ROM defined, 84
Reliability of ICs, 152
Ring counter, 79

S

Selecting ICs:
 application notes help, 153-155
 availability criteria, 150
 batch variations, 153
 choosing logic type, 149
 C/MOS generate little heat, 151
 cooling expensive, 151
 cost factor, 150-151
 critical applications, for, 153
 environmental requirements, 151
 family choice, 152
 fan-in, fan-out capabilities, 149
 for small production, 150
 frequency increases heating, 151
 hermetically sealed ICs, 152
 high reliability, 152
 important characteristics, 148
 interfacing compatibility, 149
 low speed, 149
 many specifications, 152
 MIL specifications, 152
 NAND, NOR vs. AND, OR, 150
 noise immunity, 151
 noise in C/MOS circuits, 151
 non-critical applications, for, 153
 package dissipation, 152
 packaging, 151-152
 performance requirements, 148-149
 placing performance limits, 150
 plastic packages, 151-152
 power dissipation, 151
 power requirements, 151
 regulated power source, 151
 same type number fallacy, 150
 saving power, 150
 sockets expense, 151
 special testing, 153
 specifications, important, 152
 speed defined, 149

Selecting ICs: *(cont.)*

    speed requirements, 149
    testing costly, 153
    type variations, 152-153
    variations of same type, 152-153
    varying characteristics, 153
Sense line, 84-85
Serial counters, 74-78
Scale of integration:
    defining SSI, MSI, and LSI, 26
Schmitt trigger edge generator, 108-109
Shift register, flip-flop form, 78-79
Simplifying by charting (*see* Charting to
    simplify)
Single inversion charting, 135
Small production ICs, 150
Small scale integration, defined, 26
Sockets for ICs, expense, 151
Sound level indicator, hobby project:
    advantages of, 204-205
    block diagram, 205
    colored light indicators, 204
    operational details, 205-206
    original idea, 204
Specifying ICs (*see* Selecting ICs)
Specifications, testing to:
    comparison tester, 170-172
    complete evaluation, 167
    current input important, 165
    dc test circuits, 167-170
    dc testing, 167-170
    different for digital ICs, 165
    general remarks, 165
    how to, 165-170
    input current tests, 168-169
    input voltage tests, 167-168
    limits used, 166-167
    output short circuit current, 168-170
    simple comparison test, 170-172
    supply current test, 170
    system dissipation, 166
    typical, meaning of, 166-167
    typical specification list, 166
Speed of ICs defined, 149
Square wave repeater, 63

T

Testing ICs:
    actual vs. specified data plotted, 158
    analog tests do not apply, 156

Testing ICs: *(cont.)*

    analyzer clocking, 176-177
    analyzer uses, 184
    automatic test equipment (ATE),
      185-187
    automatic testers, 184-187
    automatic testing uses, 184-185
    automatic test systems (ATS), 185-187
    box with two DIP sockets, 162
    circuit of test probe, 162
    circuits of two test boxes, 163-164
    clocking data, 160-161
    comparison, 173
    complex unit testing, 160
    digital analyzer, 160-165
    digital testing different, 156
    dual beam CRO, 160
    frequency divider testing, 160
    functions requiring testing, 156
    general remarks about, 156-157
    graphical presentation, 158-161
    high speed testing, 158-160
    high vs. low speed, 156
    individual ICs, 162-165
    infrequent testing, 161
    J-K flip-flop test, 162-165
    limit testing, 156-157
    linearity has no meaning, 156
    logic analyzer operation, 177-179
    logic analyzers, 176-184
    logic clips, 173-175
    logic probes, 172-173
    logic state analyzers, 178-184
    long digital sequence, 179
    loose specifications, 157
    memory use in analyzer, 179
    most important tests, 156
    NAND gate actual data, 157-158
    on/off criteria, 156
    parallel data analysis, 178-184
    power supply for, 162
    probe type tester, 161
    progression of signals, 160
    pulse readout, 161
    pulse response, 158-161
    rejection reasons, 157
    sequence capture, 179
    sequence testing, 176-184
    serial data transmissions, 176
    simple combination testing, 162-165
      (*see* Specifications, testing to)

Testing ICs: *(cont.)*

specified vs. measured characteristics, 157-158
spike detection, 177-178
timing with clock pulses, 160-161
triggering analyzer, 177
visual readout of pulses, 161
Three state gates, 58-59
Timer, versatile, 110-112
TTL noise margin, 115-116
Truth tables *(see* Charting to simplify)
Truth table:
AND gate, for, 31
AND/NAND, OR/NOR compared, 39
chart form of, 130-135
defined, 31
eight basic gates, 41
exclusive-OR, 40
full-adder, 91
half-adder, 90
NAND gate, 37, 41
NOR derived from OR, 32
NOR gate, 37, 41
NOT gate, 35
R-S flip-flop, 63
R-S flip-flop as latch, 64
three-state gate, 59
use in charting, 130-135

U

Understanding (transistors) in ICs, 19-22

V

Variations within an IC type group, 152-153
Voice controlled hobby device:
analyzing the problem, 193
available parts problem, 196
early block diagram, 197
final circuit, 203
final steps, 202-204
first experiments, 196-197
further developments, 200-201
generating pulses from voice, 193-194
getting an idea, 193
improved model, 199
incorporating timer, 196-197
initial tests, 193-194
input circuit testing, 193-195
output decoding, 201
shift register added, 196
timing diagram, 198

W

Watches *(see* Electronic watches)
Wired AND/OR gates, 59-60

X

X line, 84-85

Y

Y line, 84-85